Hero or Villain?

Hero or Villain?

Essays on Dark Protagonists of Television

Edited by ABIGAIL G. SCHEG
and TAMARA GIRARDI

McFarland & Company, Inc., Publishers
Jefferson, North Carolina

ALSO EDITED BY ABIGAIL G. SCHEG

*Bullying in Popular Culture: Essays
on Film, Television and Novels* (McFarland, 2015)

ISBN 978-1-4766-6769-0 (softcover : acid free paper) ∞
ISBN 978-1-4766-3052-6 (ebook)

LIBRARY OF CONGRESS CATALOGUING-IN-PUBLICATION DATA

BRITISH LIBRARY CATALOGUING DATA ARE AVAILABLE

© 2017 Abigail G. Scheg and Tamara Giardi. All rights reserved

*No part of this book may be reproduced or transmitted in any form
or by any means, electronic or mechanical, including photocopying
or recording, or by any information storage and retrieval system,
without permission in writing from the publisher.*

Front cover photography by Ivan Bliznetsov (iStock)

Printed in the United States of America

*McFarland & Company, Inc., Publishers
Box 611, Jefferson, North Carolina 28640
www.mcfarlandpub.com*

Table of Contents

Introduction   1
    ABIGAIL G. SCHEG *and* TAMARA GIRARDI

Bad Cops and Good Inmates: Shifting Power Structures   7
    in Prison Dramas *Oz* and *Orange Is the New Black*
    STEPHANIE LIM

Harrison Wells and the Making of Identities: The Scientist,   22
    the Father and the Family in the Cosmic Battle of Good
    and Evil in *The Flash*
    HANNAH SWAMIDOSS

Bad Men: The Fan Culture and Postmasculinity of *Breaking Bad*   33
    JACK CLARKE

Breaking It Down: Expressing Black Selfhood and Subverting   48
    Binaries of Good and Bad in *Key & Peele*
    HILARIE ASHTON

Talk Bluntly and Carry a Pointy Stick: Violence and Verbal   60
    Complexity in *Buffy the Vampire Slayer*
    TIRZA I. LEADER *and* DARCY MULLEN

"I need an antiheroine": Female Antiheroes in American Quality   74
    Television
    SOTIRIS PETRIDIS

A Dangerous Mind: An Examination of the Effectiveness   84
    of Walter White as Educator
    RICHARD L. MEHRENBERG

Table of Contents

Building (an) Empire: Queer Sacrifices in Lee Daniels' *Empire*   96
    ROBERT LARUE

Merlin: Magician, Man and Manipulator in *Camelot*   109
    CAROLINE WOMACK

"This isn't a democracy anymore": *The Walking Dead*'s   123
Rick Grimes and the Post-Apocalyptic Good Cop/Bad Cop
    ANNETTE SCHIMMELPFENNIG

"The girl needs a little monster in her man": Heroes and Villains   136
in the Works of Joss Whedon
    DON TRESCA

The Detective, the Pastor and the Thief: *Backstrom*'s   149
Negotiations of Morality
    ALISSA BURGER

Building and Breaking an Antihero: The Rise of Sonny   162
Corinthos
    JACINTA YANDERS

You Can't Go Home Again: The Multiple Heroisms of Sergeant   174
Nicholas Brody
    LLOYD ISAAC VAYO

Good Bad Boys and the Women Who Love Them: Romantic   186
Triangulation and the Ideal of Conformist Assimilation in
The Vampire Diaries and *True Blood*
    ANA G. GAL

Deconstructing the Dichotomy That Is *House, M.D.*:   198
A Carnivalesque Look at a Good Diagnostician/Bad Guy
    GILLIAN COLLIE

About the Contributors   211

Index   213

Introduction

ABIGAIL G. SCHEG *and* TAMARA GIRARDI

Abigail: We are big fans of *The Walking Dead* in my house. We have the comics, watch the show, lament the wait between seasons, follow up each episode with *Talking Dead*, and even *tried* to watch *Fear the Walking Dead*. I joke to my family that I would watch this cast of characters even after the zombie apocalypse ended. Once it's all over, and they're all living happily ever after, sipping piña coladas on a beach somewhere, I'd still watch them just live their lives in peace and harmony. For as much as I enjoy the characters and care about the show, though, these characters are, in fact, kind of horrible people. Sure, their environment and surroundings have forced them to do some terrible things, but other times they just chose to do the bad thing.

With the introduction of the character Negan in the season six finale, viewers finally got to see *the worst character* to ever grace the pages of *The Walking Dead* comics come to life. Following that episode, the AMC show *Talking Dead* (a post-show interview and discussion panel hosted by Chris Hardwick) welcomed Jeffrey Dean Morgan (who plays Negan), Norman Reedus (Daryl), producer Scott Gimple, and the creator of *The Walking Dead*, Robert Kirkman. Morgan was asked why he thinks Negan, who is a horrible person, has such a following.

> I think … people are also a little bit scared of him. But in that world … in that *Walking Dead* world, people are gonna follow a guy like that. Just like Rick! Rick's a little bit psychotic…. I think! I look at him and I think, "He does some crazy stuff" over the past seven years, or six years … and I think they're just two sides of the same coin [*Talking Dead*, "Last Day on Earth"].

Later, someone agreed with Morgan's assertions and argued that if we (the viewers) would have seen Negan wake from a coma to start the show (instead of Rick) and we saw how his character developed, we would like him and empathize with him. At first, I was horrified by these comments, but the more

I thought about it, the more I agreed with that assertion. I only associate Negan as a "bad guy" because he represents a foil to my "good guy," Rick Grimes. And as much as I love Rick Grimes, he blends the dichotomy of good/bad. It was in considering my love for *The Walking Dead* that my interest in developing this collection began.

I think back to the television of my youth, shows like *Saved by the Bell*, *Family Matters*, *Boy Meets World*, and *Full House*. The characters, plots, and stories were mostly good, wholesome, and family-oriented. Mistakes were made along the way, and the negative influences often came from clichés, like that one friend who was always a bad egg, or the compulsory episode about drug use or drinking and driving that also served as a public service announcement to young viewers. The people you were supposed to root for, the good characters, were clearly defined and typically didn't even associate, or only associated in a limited capacity, with the bad characters. I knew what I had to do; I had to learn more about these challenging characters.

Tamara: One afternoon when I had a stack of essays to edit for this collection, I found myself instead opening my web browser and searching for Damon Salvatore scenes on YouTube. Damon has that effect on me. For those of you deprived of the joy, Damon Salvatore is the darker half of *The Vampire Diaries* sibling duo. Played by Ian Somerhalder, Damon embodies the conflicted antihero, and my adoration of his character coupled with my coeditor's affinity for *The Walking Dead* led to a debate about why we love these good guys/bad guys so deeply. As Jacinta Yanders notes in her essay, soap operas of the 1980s regularly paired a good girl from an upper crust family with an irresistible bad boy, so the attraction of the bad boy is certainly not new to film and television. In fact, I grew up in the decade that celebrated the trope, which could have very well primed me for loving my bad boy Damon Salvatore as I do. Furthermore, as this collection demonstrates, bad guys are no longer on the periphery; they're the antiheroes, the protagonists that don't fit the mold.

Damon is one such antihero. In the pilot of *The Vampire Diaries*, Damon's younger brother, Stefan, arrives in Mystic Falls as the hot transfer student; the tortured vampire enrolls in school because of his attraction to the girl-next-door character, Elena Gilbert. Stefan and Elena's relationship shows promise, but then Damon arrives and immediately tempts Stefan to abandon his animal-feeding ways and instead taste Elena's blood. Damon's ability to antagonize his younger brother so intensely that Stefan launches them both from a second story window demonstrates Damon as the dark half of the duo. Yet Damon's "bad guy" nature softens around Elena, if only sporadically. Illustrating the good-girl-changes-bad-boy trope, Damon's respect and eventual love for Elena inspire him to choose "good" over "bad," yet he doesn't fully embrace an identity aligning with "good guys." For instance, in season three, Damon does a kind deed for his friend/lover Rose, and when Elena learns about it later, she asks why he never revealed the deed and, more generally, why he never lets people "see the good" in him. "Because when people see good, they expect good. And I don't want to have to live up to anyone's expectations," Damon responds ("Heart of Darkness"). Only moments later, Damon returns to his "bad boy" nature, tempting Elena's attraction to him. She tries to walk away, and says, "Don't," to which he replies, "Why not?" Despite her good girl core, Elena shrugs away her reservations and her guilt for feeling the way she does for her boyfriend's brother and jumps into Damon's arms for a passionate kiss that had been a long time coming and had fans sighing in utter joy. So the question becomes, what has audiences rooting for the "worthy" character to be cast aside, so the bad boy gets the girl?

Over *The Vampire Diaries*' previous seven seasons, Damon has demonstrated selfishness, brutality, jealousy, and so many other traits audiences might historically assign to the villain or antagonist, yet it is perhaps through his brother Stefan's "good guy" nature *and* simultaneous adoration for Damon that we wonder if Damon is worth saving, or maybe loving. Couple Stefan's love for his brother with good girl Elena's kindness-turned-love for Damon, and we connect with the complex character and desire his redemption while also fearing his natural inclination to destroy his life and the lives of those around him. The complexity of such characters, then, deepens audiences' relationships with the shows they love and creates a vibrant discussion for scholars debating the transformation in character roles and the good and evil that, as has been believed for some time, lives in all of us.

Lost in the stories of our respective "good guy/bad guys"—Rick and Damon—my coeditor and I debated and theorized with passion about the antiheroes, and thought other scholars might find interest in doing the same. The overwhelming response to our call for essays for this collection demonstrated we were right. There's so much value in looking at someone you believe to be wrong and "bad" yet then finding some mutual ground with the ideals that person, or character, possesses; such is true not only for fictional tales but also for the world in which we live. Or maybe better still, there is the perhaps terrifying opportunity to examine ourselves to identify the levels to which we might stoop or the ways in which our values might be compromised if our loved ones found themselves in the situations our antiheroes experience. What does exploring the stories of these characters tell us about our own selves and the lengths to which we might go to protect those we love? The essays in this collection help our audience begin to answer that question.

Abigail and Tamara: In recent years, television shows have grown tremendously in scope, graphics, and viewer expectations. To retain viewers a show must have mystery, intrigue, and, typically, plot twists no one ever saw coming. Additionally, viewers are looking for relatable characters, but characters that have many layers, and react believably in their unique fictional situations. Character arcs have become a household conversation with shows like *Breaking Bad, The Walking Dead, The Vampire Diaries, True Blood, Buffy the Vampire Slayer* and so on, which constantly muddy the waters between heroes and antiheroes. As a result, viewers might not be quite sure who the protagonist is, or the antagonist, or if maybe, due to the arcs, their beloved characters are both a little good and a little bad.

The contributors to this collection discuss this shift in popular television. They explore the changes to television in terms of complex characters, fluid character arcs, and long-lasting cultural impacts of the shows analyzed. To these authors, and to American society, the characters and the shows come alive and off the screen; the characters and shows are fodder for print publications, dinner conversations, and mass marketing and merchandising. These good guy/bad guy characters are multidimensional: complex, realistic, and challenging. For these reasons, it is necessary for a collection of this scope to explore the development of popular television characters from their roles as heroes, antiheroes, those who flip flop, and those who just can't be defined, but we love nonetheless.

Stephanie Lim, in "Bad Cops and Good Inmates: Shifting Power Structures in Prison Dramas *Oz* and *Orange Is the New Black*," uses the development of popular television's prison dramas to complicate the roles of protagonists and antagonists. Lim demonstrates that although the prisoners are clearly shown as the "bad" characters, prison staff may, in fact, have worse behaviors, but know how to manipulate the system.

In "Harrison Wells and the Making of Identities: The Scientist, the Father and the Family in the Cosmic Battle of Good and Evil in *The Flash*," Hannah Swamidoss examines the questions raised by the character Wells through the contexts of technology and knowledge and the concepts of family and community. Does the formation of a good character such as Allen, for instance, require the adverse evil catalyst of Wells?

"Bad Men: The Fan Culture and Postmasculinity of *Breaking Bad*" by Jack Clarke discusses the "Bad Man" character and demonstrates his far-reaching popularity in culture by examining online conversations and fan creations based on a love of, and reactions to, the popular television show *Breaking Bad*.

Hilarie Ashton explores the television show *Key & Peele* in "Breaking It Down: Expressing Black Selfhood and Subverting Binaries of Good and Bad in *Key & Peele*." She asserts that Keegan-Michael Key and Jordan Peele are more than comedians: they are social commentators and useful challengers to a frighteningly prevalent notion of American society as "post-racial."

"Talk Bluntly and Carry a Pointy Stick: Violence and Verbal Complexity in *Buffy the Vampire Slayer*" by Tirza I. Leader and Darcy Mullen explores the development of characters' language in *Buffy the Vampire Slayer* as they approach, involve in, and walk away from violent interactions, and, as a result, how that identifies the character as good and/or bad.

In "'I need an antiheroine': Female Antiheroes in American Quality Television," Sotiris Petridis scrutinizes the genre of "quality television" to examine some leading female characters from the popular television shows *Homeland*, *House of Cards*, and *How to Get Away with Murder*.

Richard L. Mehrenberg's "A Dangerous Mind: An Examination of the Effectiveness of Walter White as Educator" juxtaposes Walter's roles as educator and drug manufacturer/distributor to better understand the *Breaking Bad* character's effectiveness and desires to be good/bad.

In "Building (an) Empire: Queer Sacrifices in Lee Daniels' *Empire*," Robert LaRue examines family and individual dynamics in the television series *Empire*, particularly the relationship between the father Lucious and his children as they grapple with their own senses of identity. In particular, Lucious is challenged by having a queer son, based on his own upbringing and positioning within his African American community.

Caroline Womack, in "Merlin: Magician, Man and Manipulator in *Camelot*," identifies the various roles that Merlin from *Camelot* takes on during the course of the show, thus constructing his dual good and bad personas.

"'This isn't a democracy anymore': *The Walking Dead*'s Rick Grimes and the Post-Apocalyptic Good Cop/Bad Cop" by Annette Schimmelpfennig critically discusses police officer Rick Grimes as he takes on different characteristics based on the individuals that surround him.

"'The girl needs a little monster in her man': Heroes and Villains in the Works of Joss Whedon," by Don Tresca, surveys Joss Whedon's numerous popular works to better understand primary characters that reject the hero/antihero binary and consistently maneuver in the grey area in between.

Alissa Burger's "The Detective, the Pastor and the Thief: *Backstrom*'s Negotiations of Morality" scrutinizes Backstrom, a law enforcement officer. Though Backstrom often makes illegal or immoral decisions he is oftentimes an effective detective in mysterious circumstances.

"Building and Breaking an Antihero: The Rise of Sonny Corinthos" by Jacinta Yanders discusses the intrigue of the character Sonny Corinthos from *General Hospital*. Yanders uses the lasting power of soap operas to contextualize the conversation and present a historical view on the topic.

Lloyd Isaac Vayo, in "You Can't Go Home Again: The Multiple Heroisms of Sergeant Nicholas Brody," analyzes the development in *Homeland* of

Brody's character as he transitions from an Al Qaeda captive to an assumed American hero.

In "Good Bad Boys and the Women Who Love Them: Romantic Triangulation and the Ideal of Conformist Assimilation in *The Vampire Diaries* and *True Blood*," Ana G. Gal examines the consumerist ideal in these popular vampire shows. She uses the love triangles between the main characters to demonstrate that a mostly female audience is coerced into rooting for a "bad guy."

Gillian Collie's "Deconstructing the Dichotomy That Is *House, M.D.*: A Carnivalesque Look at a Good Diagnostician/Bad Guy" uses Bakhtin's theory of carnivalesque communication to examine Dr. House, whose character very distinctly plays the role of good physician/bad guy.

Collectively, these sixteen essays interrogate the concept of the antihero, or, more specifically, the complexity of a single character that could, at times, demonstrate a good guy dynamic while at other times embracing a bad guy nature. The television shows and characters analyzed have won numerous awards in recent decades, and audiences have embraced and devoured them.

Works Cited

"Heart of Darkness." *The Vampire Diaries*. Dir. Chris Grismer. The CW, 19 Apr. 2012. Television.
"Last Day on Earth." *Talking Dead*. AMC, 3 Apr. 2016. Television.
"Pilot." *The Vampire Diaries*. Dir. Marcos Siega. The CW, 10 Sept. 2009. Television.

Bad Cops and Good Inmates
Shifting Power Structures in Prison Dramas Oz *and* Orange Is the New Black

STEPHANIE LIM

The argument that those in power—law enforcement in particular—can be just as bad, if not worse, than those labeled "criminals" is increasingly being emphasized in popular American television shows. Of course, as society's view towards "good guys" changes, so does the view towards "bad guys." Recent series like *The Americans*, *Gang Related*, *Sons of Anarchy*, and *Scandal* all contain characters automatically deemed by society as "good guys" but who, more often than not, act much more evil and corrupt than the "bad guys." In fact, during what is being called a "new era" of television (Martin), Priska Neely argues, "From *The Wire* to *Boardwalk Empire*, *Breaking Bad* to the new Netflix show *House of Cards*, we've been reprogrammed to root for bad guys." Audiences witness a multitude of police officers, lawyers, and state and government officials who are supposed to be "good" commit treason, battery, and murder, while gang members, bikers, and drug dealers who are supposed to be "bad" become America's modern Machiavellian (anti-)heroes, demonstrating loyalty, love, and justice. On the surface, the country's obsession with bad guys can be associated with the fact that "we want to see the people like us, the people with flaws and mixed morals" (Martin). However, on a much deeper level, America's fascination with this upside-down dichotomy reflects societal changes and fears in the midst of never-ending stories of corruption and scandals committed by public officials and, especially in 2015, an increasing number of police brutality cases, all begging the question: what really makes a person "good" and what really makes a person "bad"?

HBO's *Oz* (1997–2003) and Netflix's *Orange Is the New Black* (2013–present) are two prison dramas that effectively blur the good guy/bad guy dichotomy—even completely reverse it at times—within the internal workings of the prison system. As both shows follow the journey of a naïve and innocent new inmate's transition into the tough and violent atmosphere of prison, audiences end up sympathizing with the inmates, seen as human beings who have made bad choices, rather than the staff, seen as relentlessly unethical, selfish, arrogant, and, disturbingly, law-breakers themselves. The shows thus illustrate how good guy and bad guy labels are instinctually applied by society simply because of a person's social status; those in power, public officials sworn to act in the best interest of the average citizen, are automatically deemed good, while those who break the law, often shown as coming from low-income neighborhoods, are automatically deemed bad, despite what the actual intents and reasons may have been. By putting the traditional "good guys" and "bad guys" in close and constant proximity to each other in a prison environment and by using elements like complex characters, fish-out-of-water scenarios, and flashbacks, *Oz* and *OITNB* demonstrate that what makes a person good or bad is not their occupation or position in society but, rather, what kind of moral and ethical compass they have.

Shifting Power Structures in Pop Culture

Within the larger umbrella genre of American crime dramas, detective shows like *Miami Vice, Nash Bridges, Castle, Rizzoli & Isles,* and *Law & Order* have clearly demarcated the difference between the good guys and the bad guys for decades; although the main "good guy" characters are often flawed (the character of Nash Bridges struggles with damaged relationships, Elliot Stabler on *Law & Order: SVU* has an extremely destructive temper, and Kate Beckett in *Castle* is haunted by her mother's murder case), the power structures in these shows are clear: the detectives are "good" because they fight the crimes, put criminals behind bars, and solve cases involving heinous crimes like murder, rape, burglary, and kidnapping. The detectives—and, by extension, law enforcement and the legal system as well—are portrayed as society's heroes. Criminals, unsurprisingly, are "bad" because they commit crimes, automatically suggesting that they are on the other side of the law, and thus, deserve legal punishments for their behavior.

Many recent cultural texts, however, have begun to express the opposite idea, that criminals may actually be good people who act with what they believe to be moral reasons. In *How to Get Away with Murder*, one character remarks, "when good people do bad things, it's usually for a reason" ("Meet Bonnie"). Such reasons exist in *Les Misérables*—Jean Valjean is branded a criminal for stealing a loaf of bread, when in fact he was trying to help feed his sister's child—and *Breaking Bad*—Walter White gets involved in drug dealing to support his family after being diagnosed with cancer. Similarly, through flashbacks in *OITNB*, audiences discover that some of the inmates are in prison for what are actually self-sacrificing or honorable reasons: characters Sister Ingalls and Brook Soso are in prison because of their social and political activism, Aleida Diaz takes a drug conviction for her boyfriend, and Claudette Pelage (known as Miss Claudette) murders a man who was physically abusing one of her workers.

In the same vein, other television shows demonstrate how good guys may actually use their power and influence to do evil. Viewers of *Scandal* continually witness the president's staff, or the president himself, commit treason and even murder, and in *How to Get Away with Murder*, a defense attorney and her group of law students use their knowledge of the law and the legal system to do everything they can to cover up a murder. Likewise, in *Oz*, Warden Leo Glynn often acts out of anger or disdain, punishing inmates for racist and personal reasons, and numerous corrections officers (COs) are seen smuggling drugs into the prisons. Even more to the point, in the flashback for *Oz*'s narrator Augustus Hill, audiences watch as the police take advantage of their power: after realizing that Hill has shot and killed a police officer, another officer takes the law into his own hands and throws Hill off the roof of a building, paralyzing him, while the rest of the officers observe and do nothing.

Irrespective of a prison's official power structure, consisting mainly of wardens, counselors, and corrections officers, the inmates in *Oz* and *OITNB* establish their own self-regulated power structures. There are powerful leaders of each racial group who others must look to for approval and guidance; there are blood laws, wherein people must answer for their wrongdoings; and the law of "quid pro quo" exists, where goods and favors are continuously bartered and exchanged, and even prison staff know how to strike deals to their advantage. Prison staff are, in fact, often coerced into playing by the inmates' rules. After a riot breaks out in *Oz*, the prison staff must abide by the inmates' demands, and after CO Bennett gets an inmate pregnant in *OITNB*, he is intimidated into smuggling contraband into the prison for the Spanish group. What become innumerable moments of weakness and powerlessness on the part of the prison staff in both series reveal a colossal shift, or even a complete reversal, of the official and traditional power relationship between prison officers and inmates.

Ultimately, the traditional power structures that would normally hold together the penitentiaries in *Oz* and *OITNB* are disrupted: prison officials are not always good people, and criminals are not necessarily bad people. With the growing number of the country's police brutality cases, particularly in light of the "Black Lives Matter" movement, *Oz* and *OITNB* ultimately function as a reflection of and reaction to the growing discontent and tension between "the law" and the civilians in the real world, making the impact and implications of the shows even more powerful today.

Oz: "Oswald Maximum Security Penitentiary—Emerald City Unit"

Throughout the first season of *Oz*, which debuted in 1997, viewers follow new inmate Tobias Beecher's transition from life as a successful lawyer into life as a convicted felon inside Oswald Maximum Security Penitentiary's Emerald City Unit, nicknamed Em City. Beecher, a hard-working family man, struggles at first to adjust to the inmate code of conduct and to his new life in prison; seen by other inmates as naïve and apprehensive with no street smarts, Beecher is marked as Aryan leader Vernon Schillinger's "prag," or "prison fag" ("Prag") and is branded with a swastika on his buttocks. Tested by disturbing experiences, including witnessing murders and corruption throughout the prison, struggling to keep his marriage and family intact, and having to submit to Schillinger, Beecher eventually becomes a violent and hot-tempered criminal himself, confirmed best by the fact that he brutally beats Schillinger and calls Schillinger his prag by the end of the first season. Beecher's ability to fight his way to the top earns him respect from other inmates and provides viewers with a lens which to evaluate the actions of the other characters. Although Beecher and other inmates struggle continuously to make moral and ethical decisions, spending time in Em City no doubt hardens each one.

The show, however, makes a substantial argument that those labeled "good guys" are not as "good" as society wants to think. Using Beecher's former social position as a lawyer and his subsequent transition into Em City as a figurative measuring stick, the series presents the corruption that exists among prison staff, the particular moral and ethical code that the inmates (and prison staff) must live by in order to survive, and the eventual reversal of power structures. To better gauge the degree of corruption among certain prison staff, there are also several staff members depicted mainly as honorable people, as they are persistently empathetic towards the inmates and fight for the inmates' rights. Tim McManus, who is the founder and Unit Manager of Em City, is a constant reminder both to his coworkers and to viewers that inmates should be viewed as human beings rather than simply as criminals. In the first episode, McManus explains to an inmate, "Do you know why in Em City, I put lifers in with all the rest? So that people can learn to live together, and not just for when they get released. Even if you're in here for the rest of your life, even if you're in here until you die, your life can have some purpose" ("The Routine"). Having created Em City because he believes that prisoners can be good people who have made bad decisions, McManus exhibits a fatherly attitude towards inmates, speaking to them with compassion and encouragement.

Similar to McManus are Sister Peter Marie Reimondo, Father Ray Mukada, and Dr. Gloria Nathan, all of whom continually fight for the inmates' rights throughout the series, at times coming head-to-head (and fist-to-fist) with the rest of the staff. Sister Reimondo and Father Mukada disagree with the prison's attitude towards senior citizens, arguing that elderly inmate Ricardo Alvarez, who is serving a life sentence in Em City, has developed "criminal menopause" and is no longer a threat to society since he does not even know where he is. To the dismay of their colleagues, they propose that Alvarez be set free, asserting that money is being wasted when the prison keeps senior citizens locked up, as older individuals who have physically and mentally deteriorated can no longer commit the original crimes they were imprisoned for. In the same way that Reimondo and Mukada attempt to reason with the higher-ups, Dr. Nathan also pushes for the full removal of a tumor, rather than a partial lumpectomy, when inmate Ryan O'Reilly is diagnosed with breast cancer. Dr. Nathan's wishes demonstrate a sincere desire and conviction to care for and look after the health of inmates, despite the violent crimes they may have committed. Unlike their cynical coworkers, civilian prison staff like Dr. Nathan, Father Mukada, and Sister Reimondo are able to see past the violence and brutality of the prisoners, instead displaying care and concern towards the inmates and treating them as human beings, even though many of Em City's occupants are indeed troubled and aggressive individuals.

The honorable goals and ideals of people like McManus and Dr. Nathan are normally shown in extreme opposition to officials in higher positions than them, such as Governor James Devlin and Warden Glynn. After a string of murders occur inside Em City, McManus argues that Governor Devlin is the real cause of the internal tension, telling him, "You banned smoking, you banned conjugals. Bit by bit you're stripping these men of their basic human needs," to which Devlin simply responds, "This is a prison. These men are criminals. The whole point is to strip them of their basic human needs" ("God's Chillin'"). Also in opposition to McManus' optimistic attitude towards the inmates is Glynn, who reminds McManus that an inmate "killed a couple on their wedding day. Shot them both down in cold blood." When McManus expounds that the couple "ripped off his package," Glynn asks, "And that justifies it?" to which McManus confidently responds, "No. It explains it," making clear that McManus believes inmates act for reasons that merit understanding ("Visits, Conjugal and Otherwise"). While Warden Glynn and Governor Devlin see inmates as a sequence of letters and numbers, prison staff like McManus attempt to connect with each inmate individually.

12   Hero or Villain?

Blinded by the need to maintain a positive reputation in the public eye, the Warden and Governor disregard the conditions in Em City, and prison guards engage in criminal activity behind McManus' back, all of which exposes the sheer corruption that exists among individuals who are—by title and position—supposed to be upright and moral. After three inmates are found fighting in a locked room and completely unsupervised, McManus concedes to the possibility that a guard may be involved in misconduct. Indeed, COs Mike Healy and Diane Whittlesey have involved themselves in illegal behavior, both agreeing to smuggle in contraband for inmates. Healy even assists inmate O'Reilly in executing attacks on other inmates, literally using his physical power to put one inmate, Bob Rebadow, in the hole and beating another, Jefferson Keane, as a warning that "nothing better happen to Ryan O'Reilly" ("God's Chillin'"). Multiple other COs are shown physically beating inmates as retaliation against the murder of an officer. The actions of these COs reveal that some officers can be untrustworthy and easily swayed by personal vendettas.

Although the inmates in Em City are almost all violent criminals, having committed murders or brutal attacks both outside and inside the prison, the inmates are still shown to live by a particular moral and ethical compass. Early in the first season, McManus initially promises that inmate Keane can get married, but later withdraws the offer; an angry Keane asserts that "a man who can't keep his word ain't shit with me. He ain't shit" ("Visits, Conjugal and Otherwise"), revealing that, although inmates have gone against official law, there are still unwritten codes of conduct that prisoners follow—in this particular case, trust and loyalty. Richard Valdemar notes that "[c]orrections officers control the containment and movement of inmates, but prison gangs and other security threat groups establish the inmate's code of conduct." The convict code is further emphasized by narrator Hill, who explains that "being in a gang is a lot like being in a religion. You got rules to follow, a leader to obey, and at the heart, it's about love. Love thy fellow man becomes love thy brother gangster" ("God's Chillin'"). Such values of brotherhood and camaraderie are largely expressed through Em City's Muslim group, in which leader Kareem Saïd continually preaches respect, non-violence, and patience. Saïd also tries to help new inmate Kenny Wrangler get back on the straight and narrow, passionately telling his fellow inmates, "That boy is 16 years old. He is in for 20, which means he can be out in 6. He can be back on the street by the time he is 22. He can still have a life" ("God's Chillin'"). Saïd's positive attitude is parallel to McManus' optimism toward the inmates, that people who do bad things are not necessarily bad people.

The power structure of Em City is most clearly inverted during the prison riots at the end of Season One. When the riots initially break out, Saïd shoots a bullet into the air and declares, "Let's get organized!" ("A Game of Checkers"). Subsequently, the prison staff quickly bend to the inmates petitions: the Warden allows inmates some time to create a list of demands, and a number of COs and staff members are thrown into a room together and closely guarded, becoming the inmates' hostages. Inmates from each racial group become riot leaders, organizing and distributing responsibilities amongst themselves. For example, Italian leader O'Reilly is put in charge of the front gate and barricade, Aryan leader Scott Ross is in charge of food distribution, Homeboys leader Simon Abedisi is in charge of any comings and goings, and Latinos leader Miguel Alvarez is put in charge of the hostages. As the riot leaders debate about whether to set some of their wounded hostages free, Muslim leader Saïd reminds the other leaders that they must fulfill their "part of the bargain ... because if we lie to [the staff], we become just like 'em'" ("A Game of Checkers"), and in so doing accuses the prison staff of being liars and frauds. Finally, in one of Saïd's most significant speeches, he explains to McManus why the riots began in the first place: "You see, this riot is not about getting smoking back, conjugal rights. It's not even about life in prison. It's about society taking responsibility; it's about the whole horrid judicial system. And we don't need more prisons, bigger prisons, better prisons. We need better justice" ("A Game of Checkers"). Saïd's discourse essentially accuses the prison officials and the legal system as a whole of being the bad guys, turning the good guy/bad guy dichotomy upside-down. The prison riots as a whole demonstrate the complete, though temporary, reversal of Em City's power structure.

Further blurring power structures is the title of the show itself. Series writer Tom Fontana makes clear that the title is meant to be ironic: Oswald Penitentiary is named after the then-prison commissioner Russell G. Oswald, making the show "a half-hearted homage to the Attica" New York prison riots in 1971 (White). The series, then, partly becomes a reframing and retelling of the prison riots just two decades prior to the series' debut. However, "Oz" is also a "subversive recontextualization of the fantasyland associated with the children's story and film *The Wizard of Oz*," and "both [Oswald Penitentiary and the land of Oz] are sites of disillusionment for the prisoners and for Dorothy respectively. Dorothy learns that the powerful wizard is just an ordinary man, and the prisoners in Oz learn all too quickly that they live in a world of chaos that cannot easily be controlled by their version of the wizard: the prison warden"; moreover, "a poster for the series also parodies Dorothy's line in the children's narrative," with the *Oz* series poster reading, "Oz: It's no place like home" (Beeler 120–121). It is no surprise, then, that the character of Warden Glynn is represented as an extremely flawed leader who frequently displays prejudice and unfairness with the inmates. In addition, the wizard in the children's story is nothing more than ordinary, implying that Glynn is really no better than any of the prisoners in *Oz*.

During *Oz*'s six seasons, various prison staff and higher-ups are shown as extremely corrupt and are continually put in vulnerable positions, signifying that those in power can often be both weak and wicked. Conversely, audiences realize that various inmates live by a particular moral code and that many of these codes are actually values that non-criminals hold too: family, loyalty, trust, camaraderie, and justice. In the end, however, *Oz*'s good guy/bad guy dichotomy only goes so far. Although the series attempts to humanize inmates who are normally dehumanized and made to feel and behave like animals, because of the men's violent tendencies, many of the inmates become proof that some criminals cannot change their ways. Viewers watch inmate after inmate die, with only five original inmates surviving through all six seasons, and all of the men are transferred to another prison at the end, which suggests that the bloodshed and fighting will continue, just under a different roof. Although it is indeed true that some characters want to be good at times—Alvarez is extremely passionate about making sure his baby, who is on life support, lives; O'Reilly does everything in his power to protect his half-brother, Cyril; and Saïd continually helps his fellow inmates work through personal situations—their desires to maintain an ethical, non-violent life are simply not stronger than their persistent anger issues and violent tendencies.

Orange Is the New Black (OITNB): "Litchfield Penitentiary"

Debuting 16 years after *Oz*, *OITNB* takes the good guy/bad guy dichotomy reversal one step further, attempting to erase the notions of "good" and "bad" completely, albeit in a more light-hearted and humorous fashion. Like in *Oz*, *OITNB* follows the life of a fish-out-of-water Piper Chapman, a well-off bourgeoning businesswoman who faces 15 months in Litchfield Penitentiary for helping her ex-girlfriend, Alex Vause, with international drug trafficking. Throughout the first season of the series, *OITNB* frequently exposes the corruption of prison officials, serving as a harsh criticism of the prison system. The series simultaneously humanizes the inmates, depicting them as morally-conscious individuals and thus making the argument that "criminals" can be good people who make bad decisions.

OITNB initially uses Chapman as the primary figure by which to question and complicate the ideas of "good" guys and "bad" guys, much like *Oz* does with Tobias Beecher. In fact, because of her privileged background and upbringing, and conceivably because she is in prison for a non-violent offense, she is often compared in extremes to the other inmates and, more importantly, likened to the guards, i.e., as "good." A self-described WASP (White Anglo-

Saxon Protestant), Chapman knows little about prison life except for what she reads and hears and has also lived a very comfortable life with her fiancé, Larry Bloom, prior to being incarcerated. During the first season, Inmate Counselor Sam Healy describes how Chapman's upbringing makes her different than the other inmates: "They're not like you and me. They're less reasonable. Less educated," and Healy later admits to Chapman that he thought they could be friends ("The Chickening"), all of which indicates that the prison staff and Chapman are thought to be alike and on the same level. Healy's statements also imply that the other inmates are uneducated, stubborn, and therefore "bad." Similarly, CO Susan Fischer acknowledges to Chapman, "I just want you to know, that, as far as I'm concerned, you and me are the same. The only difference between us is when I made bad decisions in life, I didn't get caught" ("Blood Donut"), insinuating that COs are, in fact, capable of doing bad things just as much as the inmates are, but they have simply gotten away with it. That Healy and Fischer view Chapman akin to them subverts the assumption that all inmates are automatically "bad" and, accordingly, also questions the assumption that all prison staff are automatically "good."

OITNB further complicates the good guy/bad guy dichotomy by giving viewers the chance to experience very respectable inmates, many of whom have committed non-violent crimes. For instance, when Chapman first arrives at Litchfield and enters her sleeping quarters, inmate Lorna Morello provides her with tissues because "the first night is always hard" and a toothbrush because "they don't give you one" ("I Wasn't Ready"). Chapman's roommate Anita DeMarco also gives her a roll of toilet paper and demands that Chapman let the roommates make her bed because they "know how to do it so [they'll] pass inspection" ("I Wasn't Ready"). DeMarco goes on to elucidate for Piper some of the need-to-know prison regulations as well, such as standing still for two inmate counts and explaining that dinner always occurs afterward. Head Chef inmate Galina "Red" Reznikov also demonstrates a hospitable nature when, upon meeting Chapman, gives her a yogurt cup and asserts, "You're new. You're one of us. Consider it a gift" ("I Wasn't Ready"). The inmates' generous behavior to a complete stranger is best summed up when Morello tells Chapman that "we look out for our own," clarifying that she does not mean racial ties but that the inmates only have each other ("I Wasn't Ready"). Morello's statement, in addition to the positive actions of the other inmates, demonstrates the camaraderie that exists between prisoners and the genuine care that inmates can display.

What is perhaps one of the most significant moments in Season One that illustrates the inmates' goodness occurs after a young inmate, Tricia Miller, is found dead on prison grounds. Each of the racial "tribes" shows up to Miller's group with different types of food: Chang—often categorized with the "Others"—provides a large bag of oranges, the Latino girls brings nachos, and the Black girls show up with various commissary treats as well as an illicit bottle of alcohol. These acts not only symbolize the inmates' condolences for the loss of Miller, but they also confirm a deeper understanding that inmates can be compassionate and considerate people. Later in the same episode, an exasperated Chapman expresses to her fiancé, "They're just people, Larry. They're just women who are trying to do their best. And you made them sound like they were..." and Larry interrupts, "Criminals?" ("Tall Men with Feelings"). The tense conversation reinforces, on the one hand, society's argument that all inmates are simply bad people; but on the other, the dialogue also symbolizes Chapman's—and, as a result, the audience's—quickly-changing attitude towards her fellow inmates, implying that they are not bad people at all.

Compared to the inmates' behavior, the COs at Litchfield are often exceedingly corrupt individuals; while some act in honorable ways at times, their bad conduct often outweighs the good in most cases. For example, when Chapman's fiancé arrives at Litchfield for a visit but discovers that his name is not on the list, CO Eliqua Maxwell is in no rush to help him, instead continuing to read the newspaper before slowly picking up the phone. Similarly, CO Joel Luschek, who runs the electrical shop, tells inmates to do all the work so he can take a nap and even admits, "I say a lot of shit ... then I wanna go home" ("Imaginary Enemies"), representing the complete disregard and dismissive attitude that the guards have of their duties. Moreover, Counselor Healy's frequently asks inmate Red to help with his failing marriage, and CO John Bennet falls in love with inmate Dayanara "Daya" Diaz and gets her pregnant, all of which demonstrates the fact that prison staff are unable to keep their professional and personal lives separate. Additionally, after Miller's death, Executive Assistant to the Warden Natalie "Fig" Figueroa holds a special meeting with the Litchfield staff, during which several COs make jokes in light of inmate Miller's death; even Counselor Healy does not take the meeting seriously, attempting to leave before it is over, again revealing that prison staff do not know how to conduct themselves, continuously behaving immaturely and unprofessionally.

On top of the inappropriate actions demonstrated, Litchfield's prison staff are also seen as capricious and unreliable. Healy is a primary example of such impulsive behavior, easily losing control of his own temper and of situations when under stressful circumstances; for example, Healy tells an inmate that she can take power over the television (which causes more tension between the inmates) if she would simply leave him alone. Furthermore, when Chapman voices her concern about another inmate who is threatening to kill her, Healy does nothing, leaving the area so that he cannot be held responsible for what happens. Most of all, Healy's immoral behavior is best illustrated when he creates the WAC (Women's Advisory Council) group; even though the inmates vote for the representatives they want, Healy ends up choosing the people he wants to work with, and he then bribes the inmates with donuts and coffee, asking them to forget their demands altogether. He justifies his manipulative behavior to Caputo, saying, "This isn't about giving them power. This is about your mother telling you could take a bath before dinner or after. You were still gonna get wet, but you thought you had a choice" ("WAC Pack"). Healy's reasoning illustrates how devious the prison staff can be, frequently acting on their own selfish whims and ignoring their actual responsibility which is to look after the inmates.

As more evidence of the corruption amongst the prison's officials, many of the higher-ups controlling Litchfield, like those who oversee Em City, are too concerned with the public's perception of the prison and with budget issues, continually attempting to avoid public scandals. When inmates Red and Nicky Nichols brainstorm ways to avenge Miller's death, itself a result of a guard, Red reminds Nicky that "even if they caught Pornstache stuffing drugs down [Miller's] throat, [the higher-ups would] do anything to avoid a scandal" ("Bora Bora Bora"), suggesting that those who run Litchfield will manipulate any situation if they have to. Asst. Warden Fig becomes physical proof of Red's accusations, as Fig regularly uses budget issues to dismiss the concerns of the COs and inmates. Additionally, after Pornstache is caught engaging in intercourse with inmate Daya, Fig emphasizes to fellow Asst. Warden Caputo that the situation cannot be considered rape, all in an effort to avoid a bigger scandal. In another case, after CO Bennett exposes the location from where drugs are being smuggled in, Fig tries to cover up the entire situation by finding a loophole in Bennett's story. In addition to her unethical actions, she is shown getting into a Mercedes-Benz after having just been questioned by a reporter about Litchfield's budget increase. Fig's actions and lifestyle show that the prison staff are more concerned with reputation than they are with reality.

OITNB's most noteworthy example of a corrupt staff member is CO George "Pornstache" Mendez, who relentlessly abuses his power throughout the first season. Pornstache smuggles in drugs for inmates in exchange for oral sex, and he also uses his power to take advantage of various inmates, groping Chapman during a search and completely trashing her sleeping area, damaging everything he sees, and threatening and coercing inmate Red into smuggling in drugs for him. Pornstache's biggest exploitation is covering up the death of recovering drug-addict inmate Miller; after finding that Miller has overdosed on drugs that he provided to her in the first place, he arranges the room to make it look like she hanged herself to avoid any blame. Among countless other scenarios, the horrid misconduct of Litchfield's officials, and of Pornstache in particular, clearly reveal just how repulsive the "good guys" can be.

The superficial power structure that exists in Litchfield Penitentiary

18   Hero or Villain?

encourages the prison staff, like CO Pornstache and Asst. Warden Fig, to be arrogant and blind to their own hypocrisy, as well as to the corruption of the other COs. In fact, when prison staff are seen treating inmates with any degree of dignity or respect, they are immediately criticized by their colleagues for being weak or soft, furthering the idea that inmates should be thought of as less than human. Counselor Healy, for example, has a tendency to be diplomatic with inmates in an effort to bring a sense of tranquility to Litchfield prison's atmosphere. However, Asst. Warden Caputo often mocks Healy and calls him disparaging names; for instance, when he accidentally interrupts a meeting between Healy and Chapman, Caputo jokes, "I'll come back when you're done taking orders" ("Blood Donut"), insinuating that no one actually respects Healy and that, instead, he is just being used by the inmates to get what they want. Caputo also angrily tells the COs, "You go easy on these girls and what happens? They take advantage. These women are fucking criminals. Have you forgotten that?" ("Imaginary Enemies"). Caputo's declarations strongly imply that, because the women are criminals, they will inevitably misbehave and, hence, must be kept in line. In addition, Caputo explains to CO Fischer privately:

> CAPUTO: You maintain your authority. You remind them who's in charge.... It helps if you don't use their names. Just say, "inmate," like they're all the same to you. It reminds them they're not really people.
> FISCHER: They are people.
> CAPUTO: You can't think that way. They are sheep. We feed them ... we herd them from one room to the next. They're not like you ["Fool Me Once"].

To counteract these negative connotations and assumptions about the inmates, audiences are repeatedly given the chance to empathize with individual inmates who are given humanizing and often heartbreaking backstories. Using flashbacks, viewers realize that many of the women are either in prison for doing what they believed to be morally right or for having negative external forces that influenced their decisions: Janae Watson's flashback reveals that she was a talented high school track star who felt ignored by boys, until she meets a gang member who she later commits robbery with; Piper's ex-girlfriend Vause's flashback shows her trying to reconnect with her long-lost birth father, who ends up being high on drugs and is dismissive to her, when she suddenly meets Fahri, a man who charms her into working for his drug ring; and Chapman, young and madly in love with Vause, is persuaded to help traffic drugs. Such stories are further explicated by fellow inmate Miller, who tells Chapman, "We all make bad choices. It's just some of us got different bad choices to make" ("Bora Bora Bora"). Miller's statement softens the audience's attitude towards the inmates. That is, instead of seeing the inmates as bad people who will always do bad things, audiences are encouraged to see the inmates as good people who have simply made bad choices.

The positive attitude towards inmates develops even more when inmates show remorse for the crimes they have committed. Chapman, on a phone call with her fiancé, professes, "I feel like I need to tell [Vause] that I'm sorry. I feel like I need to tell everyone ... that I'm sorry. Piper Chapman. The Apology Tour" ("WAC Pack"), depicting how regretful she is about her past actions and that her behavior has also negatively affected her friends and family. In the same way, inmate Erica "Yoga" Jones confesses to a fellow inmate that she is drunk when attempting to kill a deer and realizes that she has actually killed a young boy instead. In an emotional scene with Watson, Jones shows extreme remorse for her actions, finally shouting, "It was a mistake!" to CO Fischer ("Fool Me Once"). These scenes once again illustrate that inmates are not inherently bad but instead have just made bad decisions.

Across three seasons, *OITNB* has functioned as an explicit critique of the prison system in the real world. Healy admits to Chapman during their first meeting, "You know, I've been here for 22 years, and I still can't figure out how the system works. I got a crack dealer who is doing nine months and then I have a lady who accidentally backed into a mailman who is doing four years.... I just don't get it" ("I Wasn't Ready"). Healy's assertion exposes the flaws and contradictions in the system, in that some punishments do not fit the crimes committed and vice versa. Moreover, the series reminds viewers that the prison system can indeed be ineffective and immoral. In all, *OITNB* effectively symbolizes a reversal of the good guy/bad guy dichotomy, causing viewers to dislike the staff and instead admire the inmates.

Conclusion

Although *Oz* is nearly two decades old, the revival of an inmate-focused prison drama like *OITNB*—in light of controversial cases in America like the shootings of Michael Brown and Tamir Rice, as well as endless cases of government corruption scandals like Los Angeles County's City of Bell scandal involving the misappropriation of public funds—suggests that there are still issues with the both the justice and criminal justice systems today. All of the narrative arcs of both Em City's and Litchfield Penitentiary's inmates are constantly permeated by the outside world of race relations, gender roles, family, and the insistence of a dichotomy between good and bad, a labeling of people in order to create clear boundaries between "those who guard and those who are guarded" (Dirga and Marhánková). Moreover, by situating fish-out-of-water individuals in severe situations and by containing the stories mainly to the inner-workings of the prison system, the two prison dramas clearly serve as intricate critiques of the prison system itself. The narratives brought forth in *Oz* and *OITNB* add much to a society today ridden heavily with police brutality and government corruption scandals.

Both *Oz* and *OITNB* effectively blur the lines between good and bad, repeatedly shaking up the power relations that govern the two prisons. In *Oz*, the series not only reveals how internal power structures are created, but it also shows how inmates often have more control over what goes on in the prisons than the staff does; Em City's inmates are ruled by their own codes of conduct, fueled by values like brotherhood, loyalty, and trust, and self-regulated by rules like blood law and quid pro quo, and Em City's staff eventually act in accordance with the "con code." Although "the prisoners of 'Oz' display an almost limitless capacity for betrayal and for inflicting pain," the series is able "to find a core of humanity in each" (Smith). Likewise, in *OITNB*, the series exposes the sheer corruption of prison staff, depicting cops as disgusting but inmates as likable. As a result, audiences tend to cheer for the inmates rather than the prison officials, thereby reversing the dichotomy between good guys and bad guys. Overall, both shows offer non-essentialist views of people, presented in the fact that the labels of "staff" and "inmates" do not define who a person is, nor do they have any bearing on how a person will behave. By showing corruption from the top down, and specifically in a prison setting, *Oz* and *OITNB* demonstrate that the differences between prisoners and prison staff are not as clear as the gates dividing them.

WORKS CITED

Beeler, Karin. *Tattoos, Desire and Violence: Marks of Resistance in Literature, Film and Television*. Jefferson, NC: McFarland, 2006. Print.
"Blood Donut." *Orange Is the New Black: Season 1*. Writ. Jenji Kohan. DVD. Lionsgate, 13 May 2014.
"Bora Bora Bora." *Orange Is the New Black: Season 1*. Writ. Jenji Kohan. DVD. Lionsgate, 13 May 2014.
"The Chickening." *Orange Is the New Black: Season 1*. Writ. Jenji Kohan. Lionsgate, 13 May 2014. DVD.
Dirga, Lukáš, and Jaroslava Hasmanová Marhánková. "Nejasné Vztahy Moci—Vezení Ocima Ceských dozorcu / Prison as a Place of Ambiguous Power Relations: The Perspectives of Czech Prison Guards." Abstract. *Sociologicky Casopis* 50.1 (2014): 83–105. *ProQuest*. Web. 30 Oct. 2015.
"Fool Me Once." *Orange Is the New Black: Season 1*. Writ. Jenji Kohan. Lionsgate, 13 May 2014. DVD.
"A Game of Checkers." *Oz: The Complete Seasons 1–6*. Writ. Tom Fontana. HBO Studios, 17 Apr. 2007. DVD.
"God's Chillin'." *Oz: The Complete Seasons 1–6*. Writ. Tom Fontana. HBO Studios, 17 Apr. 2007. DVD.
"I Wasn't Ready." *Orange Is the New Black: Season 1*. Writ. Jenji Kohan. Lionsgate, 13 May 2014. DVD.
"Imaginary Enemies." *Orange Is the New Black: Season 1*. Writ. Jenji Kohan. Lionsgate, 13 May 2014. DVD.
Martin, Lauren. "Villains Do It Better: Why This Generation Is Obsessed with the Anti-Hero." *Elite Daily*. DMG Media, 23 Dec. 2013. Web. 20 Nov. 2015.

"Meet Bonnie." *How to Get Away with Murder.* ABC, 22 Oct. 2015. Television.
Neely, Priska. "'The Americans': When You're Rooting for the Bad Guy." *NPR.* NPR, 7 Feb. 2013. Web. 20 Nov. 2015.
"Prag." *The Rice University Neologisms Database.* Suzanne Kemmer, 10 June 2008. Web. 11 Sept. 2015.
"The Routine." *Oz: The Complete Seasons 1-6.* Writ. Tom Fontana. HBO Studios, 17 Apr. 2007. DVD.
Smith, Dinitia. "Prison Series Seeks to Shatter Expectations." *The New York Times.* The New York Times Company, 12 July 1999. Web. 11 Sept. 2015.
"Tall Men with Feelings." *Orange Is the New Black: Season 1.* Writ. Jenji Kohan. Lionsgate, 13 May 2014. DVD.
Valdemar, Richard. "Understanding Inmate Codes of Conduct." *POLICE Magazine.* POLICE Magazine, 25 Apr. 2011. Web. 25 Oct. 2015.
"Visits, Conjugal and Otherwise." *Oz: The Complete Seasons 1-6.* Writ. Tom Fontana. HBO Studios, 17 Apr. 2007. DVD.
"WAC Pack. *Orange Is the New Black: Season 1.* Writ. Jenji Kohan. Lionsgate, 13 May 2014. DVD.
White, Armond. "Tom Fontana: Wizard of Oz." *NY Press.* Straus News, 16 Apr. 2002. Web. 11 Sept. 2015.

Harrison Wells and the Making of Identities
The Scientist, the Father and the Family in the Cosmic Battle of Good and Evil in The Flash

HANNAH SWAMIDOSS

A disquieting moment occurs in the first season of the television series *The Flash* (Berlanti, Kreisberg, and Johns 2014–), when archvillain Eobard Thawne/Harrison Wells calmly explains to his protégé Barry Allen, the Flash, why Wells killed Barry's mother. The time-traveling Wells recounts a future enmity between himself and the Flash (both characters possess super speed and are meta-humans) in which neither can master the other. Discovering the Flash's secret identity, Wells decides to go back in time to kill Barry. The young Barry, however, is rescued, and Wells recounts his thoughts before killing Barry's mother: "What if you were to suffer a tragedy? What if you were to suffer something so horrible, so traumatic that your child self could never recover? Then, you would not become the Flash" ("Fast Enough"). What Wells does not foresee is that he will become stuck in this time, and his only means of returning to his own time is through the Flash, the very being that Wells hoped to effectively destroy. Wells tells Barry, "the Flash was gone, and so I created him" ("Fast Enough"). In creating the Flash, Wells displays a cold clinical ruthlessness against anyone who could thwart his goal. Yet, Wells also works with Barry and the police force to stop criminals and does great good in the midst of pursuing an ulterior selfish goal. To further complicate the situation, Wells offers Barry his own chance to time travel to save his mother and "undo all the evil I've done" ("Fast Enough"). Fittingly, if Barry were to take the offer, while he traveled back in time, Wells would travel forward to his own time. Until the moment he is "erased," Wells paradoxically epitomizes evil *creating good*.

Literature on superheroes frequently focuses on how these characters maintain national interests. With superhero comics first appearing in the late 1930s and continuing through World War II, superhero narratives during this time period (known as the "Golden Age") reflect both the unease and patriotism created by global events. In his assessment of the television programs depicting animated superheroes from 1978 to 2011, Kevin D. Williams notes that such shows carry on a tradition seen in the Golden Age of comic books. Williams observes, "superheroes came to represent more authoritarian beliefs. They are united in solidarity with American troops, always vanquishing evil with unrivaled superiority and fighting for 'The American Way,' which at its backbone was truth and justice" (1335).[1] In this context, Wells, the villain, seems to be an anomaly because he does not threaten national interests in his single-minded purpose to travel back to his own time. Wells' overriding desire to return, however, indicates a home lost. By having both the hero and the villain ultimately desire the same good thing—a restored home—*The Flash* positions the family at the heart of national security.[2] By having a villain who does good to further his evil purposes, the first season of *The Flash* engages with the typical elements of superhero narratives (heroes, villains, and issues of justice) in markedly new ways. Because Wells does not primarily threaten the American government, but the family, the first season of *The Flash* diverges significantly in the moral positions it takes from earlier superhero narratives. The series uses Wells' mixture of good and bad to remake the role of the scientist in society and to redefine the family. Since Wells so effectively blurs the lines between good and evil, he becomes crucial in understanding the moral stance the first season takes on the scientist and the family and their effects on society. This essay will, consequently, examine Wells in the roles of scientist and father and demonstrate that his combination of good and evil helps characters understand the larger issues at stake and configure a new moral community.

Harrison Wells the Modern Prometheus

In the narrative's portrayal of Harrison Wells—the scientist who is a God-like creator—*The Flash* continues a long-standing conversation about the role of the scientist, especially when superheroes and villains gain powers through scientific endeavors gone awry. Shortly after World War II, the role of the scientist became contested in multiple venues after key events occurred such as the bombing of Hiroshima and Nagasaki and the launch of Sputnik. Mary Shelley's *Frankenstein* (1818) also enters this discourse on the role of

24 Hero or Villain?

the scientist: Robert Genter, citing Stan Lee's autobiography, notes that the Hulk was based on *Frankenstein* (961). Shelley's depiction of Victor Frankenstein also makes an interesting contrast to Wells, and looking at some salient points in scholarship on *Frankenstein* and depictions of scientists during the Cold War prove helpful in understanding how *The Flash*'s portrayal of Wells blurs the distinct lines of good and evil and the ensuing moral issues this distortion creates.

Critics often read Dr. Frankenstein as usurping, or trying to dominate the natural order and representing a transgressive form of science.[3] With respect to the "science question," Suparna Banerjee argues that Frankenstein's failure in creating a being who has a place in society (the monster's size and appearance instantly creates fear) ties into early nineteenth-century debates about nature and culture and domestic ideology (9). Banerjee argues: "Through the disastrous results of Frankenstein's ambitious technological experiment Shelley makes the point that any attempt to treat the natural aspect of the human animal to the exclusion of the cultural is untenable, for it is through a constant interpenetration of the natural and the cultural that the human constitutes and expresses itself" (10). Unlike this crucial error that Frankenstein makes, Wells quickly learns that he needs to place the Flash culturally; Wells realizes that for his purposes of the Flash attaining maximum speed (the human animal), the Flash needs to be *morally good* (the cultural).[4] Wells not only actively helps Barry thwart evil, but Wells also creates a "work family" with two other scientists for Barry wherein he finds support that his adoptive family cannot fully provide. Yet Wells' success in the area in which Frankenstein fails only further emphasizes the difference between the moral nature of these two characters. While Frankenstein's creation of the monster is self–serving in some ways, Frankenstein does hope to help humanity through his creation. Wells, on the other hand, only seeks to benefit himself and the good that results from his efforts with the Flash is a superfluous byproduct. Ironically, Wells goes out of his way to do good to further his self-serving goals. Instead of having a well-intentioned scientist unintentionally unleashing destructive forces on humanity, *The Flash* presents the unprincipled scientist creating good to further his purposes.

When looking at the depictions of scientists in comic books written during the Cold War era in America, the complexity of Harrison Wells' character become even more apparent. Unlike Frankenstein who inadvertently releases destructive forces on the world, the scientist, in the Cold War era is seen as deliberately creating destruction, particularly through creating technology for more effective warfare. Genter convincingly argues that Stan Lee successfully rehabilitates this negative image of the scientist in the figures of Mr. Fantastic and the Hulk. Whereas Mr. Fantastic represents the family man/scientist, the Hulk in his physical prowess addresses fears about the destructive nature of science and contemporary concerns about masculinity; the Hulk's brute strength demonstrates that science's agency can be used for good, while also restoring the American male's masculinity (Genter 962–965). In contrast to Frankenstein wherein lack of social placement creates great havoc, the Hulk's lack of placement indicates a certain amount of angst, but also a tremendous amount of unrestricted masculine agency. Wells joins this discourse of placement and displacement with his trademark ambivalence. Wells has a brute aspect to his nature similar to the Hulk's: Wells (like his creation the Flash) has great speed which endows him with a tremendous amount of physical prowess and ability. More importantly, Wells has social status and placement in his community (although at times he is seen as creating destruction) but remains isolated like the Hulk which again provides agency. Yet a crucial shift occurs between the Hulk and Wells. The Hulk's isolation does not affect his moral character whereas Wells' isolation does. In his analysis of the Hulk, Genter observes that "[o]n one level, Shelley's novel is a parable about the dangers of violating the basic tenets of scientific practices predicated upon the free exchange of information" and sees the similarity with the Hulk in that "Dr. Banner refuses to share the secrets of the bomb with his fellow scientists" (961). *The Flash* makes this same point when Caitlin states: "In science we share; we do not keep secrets" ("Fastest Man Alive"). Wells definitely does not share his scientific knowledge unless it serves his purpose of making Barry faster. While Wells needs to be secretive to safeguard his plans, his knowledge of future technology could have advanced his assistants' skills or scientific knowledge within bounds.[5] *The Flash* privileges scientists working within a close network professionally over the type of isolationist position that Wells takes—even when he achieves great good. This position on the scientist reveals how far *The Flash* has diverged from the Hulk of the Cold War era. The scientist's subject formation needs to be deeply integrated into society, not set apart or elevated for the sake of national security.

With respect to the scientist, the first season of the series makes it clear that the kind of scientist one chooses to be matters a great deal. Despite his talents and the good Wells has achieved and set into motion, his extreme self-absorption leads to his "erasure," and Wells becomes a cautionary tale to the three scientists who work with him, Barry, Caitlin, and Cisco. Wells becomes crucial to their subject formation as scientists by demonstrating how easy it is to make indistinct the lines between good and evil. Yet, the series does not only engage with the scientist with respect to professional models, but also engages with the scientist at another level—the scientist as father. In spite of his deliberate isolationist tactics, Wells develops a close relationship with each of the younger scientists who work with him. While reminiscent of Mr. Fantastic's fatherly role from the Cold War era, the first season of *The Flash* presents a new form of the family in which the scientist not only commits to the scientific family, but the human family. Wells' mix of evil producing good is of great significance in this area as well.

Harrison Wells the Modern Single Father

In the same way in which the series portrays the unusual paradigm of the evil scientist creating good, the narrative also presents the uncommitted, ill-intentioned father having a positive impact on his children. By having Wells take the more intimate fatherly aspect of creator, *The Flash* enters discourses about masculinity and paternity begun in the 1980s and continuing to the present.[6] In her classic assessment of Hollywood's portrayal of fathers, Stella Bruzzi analyzes a range of father figures including the traditional father, fathers facing the complications of race or sexual orientation, reformed fathers who despite an initial rejection of fatherhood come to accept their paternal role, and bad fathers who damage or destroy their children's lives. Instead of the reformed father or the bad father, *The Flash* offers the individual manipulating his paternal role for his own benefit, yet having a positive impact on his children. To further complicate this intent, the narrative portrays Wells coming to appreciate the pleasures of being a father, but not enough to give up his plans to go home. The series uses Wells' mix of evil and good to make significant points about the family.

To understand the type of fathering that Wells represents, an examination of Barry's biological father, Henry Allen, and adoptive father, Joe West, proves useful. Imprisoned with a life sentence, Henry seems the least able to provide parental care to his son, yet despite Henry's powerlessness in prison, he holds tremendous influence over Barry. Likewise, Henry displays elements of self-sacrifice; he does not want to be freed if that were to change Barry's character and identity. Henry's integrity positions the biological father's absence in the best possible light. Yet, *The Flash* also subverts the centrality of the biological family, partly because of its fragility. Henry's full presence in Barry's life is not necessary for Barry to develop into a responsible moral being. *The Flash*'s depiction of fathers and families moves past the nuclear family. Instead of the biological father, the narrative uses the non-biological, and not legally related father Joe, to take care of eleven-year-old Barry's needs and to nurture him into adulthood.[7] The narrative's portrayal of Joe not formally adopting Barry is partly to maintain a tricky balance in the story arc of the first season in which Iris does not relate to Barry like a sister but as his best friend and love interest. Yet, Joe's lack of legal standing as Barry's father further reveals the kind of non-nuclear, non-biological family the series privileges. By not making it seem out of the ordinary that Joe would take in his neighbors' child, and by repeatedly emphasizing Joe's integrity in treating

Barry like his own son, *The Flash* makes a case for adults to address the gaps left by the nuclear, biological family.[8] Significantly, when Barry asks Joe for advice about Wells' offer to time travel and save his mother, Joe insists that Barry should take the opportunity to have his biological family intact. Yet the season ends with Barry still placed in the West household—the non-biological, legally informal parent and family becomes the *morally* correct choice.

The narrative's addition of Wells as the third father to this already established and supportive mix of fathers, consequently raises, questions as to the purpose of his paternal role. The series goes out of its way to show the amount of harm Wells has inflicted on Barry. Wells is the reason for Henry's imprisonment and Barry's mother's death; Wells damages the nuclear family and sidelines the biological father. Nevertheless, Wells can address an important aspect of Barry's life: Wells understands the scientific Barry and the *metahuman* Barry better than Henry or Joe. Being both a scientist and metahuman, Wells is the closest to Barry in identity, and Wells becomes crucial to Barry's subject formation. Wells can mentor, counsel, and encourage Barry in ways that no one else can, and Wells does the same for Caitlin and Cisco. Wells' role as father demonstrates that a child cannot have too many parents; multiple parents cover the gaps that one or two parents may leave. Since Wells addresses gaps in the family that only a scientist can fill, the first season of the series depicts the male scientist needing to commit to be a father to further *science*; without Wells, the younger scientists would have struggled to fully realize their abilities. As mentioned earlier, Wells is similar to the Cold War era's Mr. Fantastic in his fatherly interactions with the Thing and the Human Torch. A subtle shift, however, has occurred. Mr. Fantastic's co-parent is Susan Storm another scientist, while Wells' co-parents Joe (a policeman) and Henry (a doctor) are not. Wells, through familial ties, becomes closely connected to the non-scientific world. Even though Wells furthers science by nurturing scientists, the family of scientists is no longer a self-contained community benevolently helping the larger human family, but instead is deeply embedded within the human family and receives as much as it gives. As much as Barry, Caitlin, and Cisco need parenting from a scientist, they also need to be integrated with their non-scientific family members.

Wells' crucial position in the scientific and human families raises the question of whether the good produced by his effective parenting outweighs his evil intent. Wells' erasure at the end of the first season initially seems to privilege the need for good parents like Henry and Joe no matter how limited or flawed. While this type of good parent needs to be the norm, the series also explores a morally gray area in which the good Wells does is not necessarily negated by his evil. To better understand this position, a look at how Wells parents Barry, Caitlin, and Cisco is helpful. Although Wells' primary reason for parenting the younger scientists is to return to his own time, the narrative presents Wells taking care of the younger scientists' needs when he does not have to or even when it proves detrimental to his own goals. Wells, for instance, helps Barry to take risks in helping others, risks in which Barry may die, destroying Wells' plan to return to his home. Likewise, Wells decides to disrupt his schedule to return home for Caitlin's sake and uses a rare technological device on her behalf which he went to great lengths to obtain to help with his plans to return to his time. Wells truly enjoys Cisco's company and tries to help Cisco with his relationship with his biological family. Likewise, Wells repeatedly affirms Cisco's abilities and his worth to the team, even when Cisco fails to the extent of revealing Barry's real identity to a criminal. An explanation for these deviations in which Wells genuinely does things for Barry and the others occurs in the last episode of the series when Barry expresses his fury to Wells about his deception. Wells responds, "I know that rage. I used to feel that rage every time I looked upon you. And now, somehow, I know what Joe and Henry feel when they look on you with pride, with love" ("Fast Enough"). Because Wells sees the Flash before their enmity occurs and watches Barry grow up and develop into the Flash, Wells has the unexpected experience of seeing Barry from a different perspective, a fatherly perspective. Wells' identity changes from the experience of being a father to the extent that he can, at times, act on the behalf of others instead of himself. As heinous as his acts against Barry are, the narrative depicts Wells having an opportunity for redemption through fatherhood. Wells, however, does not rightly value this opportunity, and his rejection of his role in his new family will ultimately lead to his erasure.

This possibility of redemption for Wells (without minimizing the heinous nature of his acts against the human family) indicates a new understanding of the individual and the community brought in by the restructuring of the family. Wells' unique combination of evil creating good helps understand key moral principles of the type of community the first season of the series advocates. Since issues of justice in society are central to superhero narratives, the season uses the evil that Wells represents to structure parameters for punishment and redemption in its vision of a new, moral community.

A Delicate Balance: The Individual and Community

In a discussion of crime in comic books, Nickie D. Phillips and Staci Strobl analyze the choices heroes make with respect to villains, and their analysis proves useful in examining the first season's portrayal of justice and redemption. Phillips and Strobl argue that "in administering extralegal justice, heroes make 'deathworthiness' calculations along the way, deciding whether to kill villains or let them live" (8). Phillips and Strobl then note: "Rather than deathworthiness decisions being made solely on the basis of the legal culpability of the offender, we find that such decisions hinge on the intrinsic nature of the hero him/herself" (8). Applying this concept to *The Flash* reveals aspects of how good and bad are defined in the first season. Significantly, Barry, the hero, does not make these "deathworthiness" choices by himself; instead, issues of justice are debated by different characters, and team Flash soon realizes that there are many gray areas in the situations that they experience. The meta-humans who turn to criminal activities provide an example. The first two meta-human criminals stopped by the team die: one is shot when he is about to kill Barry, and the other chooses to die and falls off a roof. Joe points out, however, that a future plan needs to be devised: the team cannot "execute" every criminal meta-human. Because regular prisons cannot restrain meta-humans, the scientists build a special prison, with the intent of eventual "rehabilitation" ("Things You Can't Outrun," "Rogue Air"). All involved in this project, including Joe, realize that this solution is not the best, but it is the best that they can do in the midst of rapidly evolving situations. The ideas of imprisonment and rehabilitation (rather than execution) point to the worth of these individuals despite their criminal activities. The team takes on risks not only for themselves, but also for the community, and these risks become more apparent when the meta-human criminals escape custody. The first season's construct of a moral society is that rehabilitation should be extended to harmful individuals because of the high worth of the individual, even when the individual hurts the community. And so the possibility of Wells' redemption can occur.

Wells' erasure, however, points to the responsibility the individual has to sacrifice himself or herself for the good of the community. Wells values the individual, his own need to go home, to a faulty "deathworthy" extent. Significantly, this deathworthiness can be best seen in the type of home that Wells values. Even though Wells clearly understands the worth of the home he has lost, he does not exhibit the ability to accept the new home that he is given. Barry, on the other hand, longs for his biological family, but also genuinely loves and commits to his adoptive family. Consequently, Barry can draw others into his familial circle and sacrifice for them at the same level he would for Henry, Joe, and Iris even when the extended family members are criminals. Instead of the categories of "them" and "us," bad guys versus good guys, Barry appreciates the value of "us and us." Wells, conversely, only values "I and no one else." Characters like Barry's mom, who dies selflessly, and Henry, who bears his false imprisonment well, understand the personal sacrifice needed to maintain this model of the family, the community, and humanity. The community that ensues from such familial commitment is a deeply interdependent community. This new community finds agency in dependence (characterized by self-sacrifice and by balancing the needs of the individual and the group) instead of the self-serving independence seen in Wells. In the first season of *The Flash*, this emphasis on the relational (which builds a more inclusive community) shifts agency from superpowers and science/technology to moral sacrifice made on behalf of the community. Since every individual can invest in the family and become part of the family, everyone's set of skills is of value. This inclusiveness makes this new structure of the family always growing in a creative, original manner, and the family's expansive nature offers a place for all. Strangely, but appropriately, both Wells' good and bad sides make a case for this position.

Ultimately, by positioning the scientist in the midst of a relational model and emphasizing the need for investment in the fluid, inclusive, and expanding structure of the post-nuclear, non-biological family, *The Flash* rejects simplistic and rigid formulations of "good" and "bad" characters. With *The Flash*'s emphasis on the family as the paradigm for society, the narrative offers a fine balance between the individual and the group. When each individual sacrifices for the family and the family sacrifices for the individual, each member of the family becomes exceptionally important and a tremendous form of agency. It is this agency that ultimately defeats Wells. While the brokenness of the family may indicate an ever-present evil, it also sets the stage for great, redemptive good. The narrative's model of the family gives each individual the opportunity to participate in family life; this participation, subsequently, makes every individual a key player in the security of the nation. The paradoxical figure of Eobard Thawne/Harrison Wells, with his complicated mixed identity of good and evil, makes the strongest case for this new family structure.

Notes

1. While Robert Genter's study of Marvel comics during the Cold War period demonstrates that characters like the Hulk and Iron Man display differences or tension with the authoritarianism of the government, they too work along with the government to protect national interests. In a study of the portrayal of the criminal justice system in comics, Nickie D. Phillips and Staci Strobl observe that comic books are seen as reinforcing the status quo (8–9).

2. During the opening credits of all twenty-three episodes of the first season, for instance, Barry describes his mother's death, his father's wrongful imprisonment, and Barry's desire to gain justice for his parents.

3. Anne K. Mellor and William Patrick Day, for instance, view Frankenstein as representing the male's (and science's) will to dominate and control the natural/the feminine. Day sees Frankenstein trying to circumvent the female reproductive role, and Mellor argues that Shelley critiques Frankenstein for being antievolutionary. Matthis Krischel notes that *Frankenstein* reveals the anxiety created by treatments for nervous disorders and hysteria and brings medical ethics into question.

4. In the contexts of transgressive science and the social placement of a "created being,"

the series portrayal of Wells' creation of the Flash proves interesting. Wells clearly transgresses in certain areas—he kills the real Harrison Wells and his girlfriend and takes on the physical identity of Wells (including his DNA) through technology of the future. Yet in the actual creation of the Flash, Wells does not necessarily represent a transgressive form of science. Having traveled from the future, Wells knows that in 2020, Barry becomes the Flash through a "particle accelerator" created by the real Harrison Wells. The false Wells uses his future knowledge to "make" the Flash six years ahead of schedule. Since Wells only manipulates the timing of the Flash's creation and not the creation itself, Wells does not usurp the natural order in the manner in which Shelley's Frankenstein does.

    5. Different characters make it clear that changing events through time travel can have cataclysmic effects. Wells claims that he has made the least amount of changes possible to the time period he landed in to accomplish his plans.

    6. CBS's *Murphy Brown* (1988–1998) captures the crisis seen in the family when Brown, a single, successful professional woman decides to have a baby outside of marriage. The fictional Brown's choice becomes a matter of public debate including the 1992 presidential campaign: Dan Quayle comments on Brown "mocking the importance of fathers" (Hartman 387). In 1995, Bill Clinton calls attention to "a crucial area of responsibility, the responsibility of fatherhood" and notes that "[t]he single biggest social problem in our society may be the growing absence of fathers from their children's homes" (n. pag.). In this context of the crisis in the family Estella Tincknell argues that fathers were gaining visibility in both film and television and were being portrayed in new relational roles while women became absent or marginalized in these families (55–57). In a discussion of films, Jude Davies and Carol Smith observe: "Since the late 1980s representations of white males as domesticated, feminised or paternal have featured prominently in numerous films in a range of genres" (19). In her classic assessment of Hollywood's treatment of fathers, Stella Bruzzi notes that these new fathers uphold traditional patriarchal structures (158). Nicky Falkof sees the principle holding true in the television series *Mad Men* (2007–2015) where she argues, "we experience the father's absence as maternal rather than paternal failure" (40).

    7. *The Flash* portrays the resulting racial mix (Joe and Iris are African American, Barry Caucasian) and Joe's not having legal custody of Barry as unproblematic. In different episodes, Joe and Iris discuss some initial stress when Barry becomes part of the household but none of this stress touches upon race.

    8. The narrative portrays Joe as a tough, no-nonsense, good cop who can nurture both his children, even when he does not talk much or keeps private the uglier aspects of his work. Joe is fiercely protective of both Iris and Barry. Even after Joe sees Barry's abilities as the Flash, Joe does not want Barry involved in the pursuit of criminals until Joe realizes that with meta-human criminals Joe needs the Flash's help. Likewise, Joe does not want Iris to know that Barry is the Flash in order to protect her.

Works Cited

Banerjee, Suparna. "Home Is Where Mamma Is: Reframing the Science Question in *Frankenstein*." *Women's Studies* 40 (2011): 1–22. *EBSCO.* Web. 13 Apr. 2013.
Bruzzi, Stella. *Bringing Up Daddy: Fatherhood and Masculinity in Post-War Hollywood.* London: British Film Institute, 2005. Print.
Clinton, Bill. "Address on Race Relations." University of Texas at Austin. 16 Oct. 1995. Transcript from Miller Center, University of Virginia. Web. 15 Sept. 2015
Davis, Jude, and Carol R. Smith. *Gender, Ethnicity, and Sexuality in Contemporary American Film.* London: Routledge, 2000. Print.
Day, Walter. "Victor and His Creation Struggle with Gender Identity." Don Nardo, ed. *Readings on Frankenstein.* San Diego: Greenhaven Press, 2000. 78–83. Print.
Falkof, Nicky. "The Father, the Failure and the Self-Made Man: Masculinity in *Mad Men*." *Critical Quarterly* 54.3 (October 2012): 31–45. *EBSCO.* Web. 15 Nov. 2015.
"Fast Enough." *The Flash.* The CW Network. 19 May 2015. Television.
"Fastest Man Alive." *The Flash.* The CW Network. 14 Oct. 2014. Television.
Genter, Robert. "'With Great Power Comes Great Responsibility': Cold War Culture and the

Birth of Marvel Comics." *Journal of Popular Culture* 40.6 (2007): 953–978. *EBSCO*. Web. 6 May 2015.

Hartman, Anne. "Murphy Brown, Dan Quayle, and the American Family." Editorial. *Social Work* 37.5 (September 1992): 387–388. *EBSCO*. Web. 15 Nov. 2015.

Holloway, Daniel. "The CW: Bringing the Boys Back Home." *Broadcasting & Cable* (May 19, 2014): 15. *EBSCO*. Web. 5 Dec. 2015.

Krischel, Matthis. "Electricity in 19th Century Medicine and Mary Shelley's *Frankenstein*." *AUA/News* (January 2011): 20–21. *EBSCO*. Web. 16 Apr. 2013.

Mellor, Anne K. "*Frankenstein*, Racial Science, and the Yellow Peril." *Nineteenth- Century Contexts* 23 (2001): 1–28. *EBSCO*. Web. 16 Apr. 2013.

Phillips, Nickie D., and Staci Strobl. *Comic Book Crime: Truth, Justice, and the American Way*. New York: New York University Press, 2013. Print.

"Plastique." *The Flash*. The CW Network. 11 Nov. 2014. Television.

Tincknell, Estella. *Mediating the Family: Gender, Culture, and Representation*. New York: Oxford University Press, 2005. Print.

Williams, Kevin D. "(R)Evolution of the Television Superhero: Comparing *Superfriends* and *Justice League* in Terms of Foreign Relations." *Journal of Popular Culture* 44.6 (2011): 1333–1352. *EBSCO*. Web. 23 Sept. 2015.

Bad Men
The Fan Culture and Postmasculinity of Breaking Bad

JACK CLARKE

Since the premier of *The Sopranos* (1999–2007) on the premium cable network HBO, the past fifteen years of American quality television have been dominated by male lead protagonists who would traditionally be portrayed as villains. This "Bad Man" archetype, as it will henceforth be labeled in this essay, is typically a white, talented, middle-aged antihero who leads a double life, engages in serious criminal activities and exhibits various immoral traits and behaviors (e.g., adultery, manipulation, violence). Additionally, the Bad Man will often, directly or indirectly, raise ethical questions and challenge their audience's sense of morality, positioning viewers to "emotionally invest in, even root for, even love" these morally compromised characters that would be condemned in reality (Martin 4).

Examples of the Bad Man series currently in production include, *Billions* (2016–), *Better Call Saul* (2015–), *Homeland* (2011–), and *House of Cards* (2013–). Recent past examples include *Boardwalk Empire* (2010–2014), *Hannibal* (2013–2015), *Sons of Anarchy* (2008–2014), *Mad Men* (2007–2015), *The Shield* (2002–2008), *Dexter* (2006–2013) and *Breaking Bad* (2008–2013). Notably, all of these shows have been met with widespread critical acclaim, and more than half of them have spanned five or more seasons on air; their contemporary prestige, presence and popularity is undeniable.

In addition to this critical and commercial success, each of the aforementioned series have inspired devoted fan followings, enabling the Bad Man to enter the realm of popular culture phenomena through the active creation and maintenance of online paratexts. These fan-generated paratexts include online appreciation groups, discussion forums, tribute videos as well as the regular creation, replication and reposting of visual internet memes. As a result of this active fandom, popular Bad Man characters, scenes, or quotes are elevated to the status of pop-culture icons, becoming recognizable and celebrated images throughout online social media and, consequently, in the daily lives of the audience.

Although this "televisual pop-culture icon" status is by no means exclusive to the Bad Man, the pervasive presence of morally compromised protagonists throughout social media does illustrate the extent to which these characters have been accepted and celebrated by mainstream audiences. The purpose of this essay is to investigate how these Bad Man television texts and their audiences (as represented by their fan-related activities) are representative and reflective of contemporary American ideals of masculinity. Accordingly, this essay will take a reception studies approach in its investigation of how audiences have related and responded to *Breaking Bad*'s lead protagonist Walter White (Bryan Cranston) as a Bad Man through active participation in online social media. This essay also takes a particular interest in how these masculine-focused television series portray their supporting female characters and, in turn, how these figures are received by an audience that has been positioned to identify with the Bad Man protagonist.

* * *

In his book *Difficult Men*, Brett Martin acknowledges *The Sopranos* as a major instigator of an American "television revolution" (4) and argues that Tony Soprano set the precedent for the various masculine criminal protagonists of today, labeling the recent prominence of the Bad Man as "a new Golden Age" in television (9). Martin then goes on to trace the influences of the Bad Man archetype throughout the history of American cinema, citing Western and gangster movies as clear precursors for the Bad Man series. Martin also observes the influence of the outlaw protagonists featured in the seminal New Hollywood texts of the late 1960s and early '70s, including *Bonnie and Clyde* (1967), *Easy Rider* (1969) and *The Godfather* (1972) (84).

In addition to these broad, earlier influences, *Scarface's* (1983) villainous protagonist Tony Montana can arguably be observed as the most direct precursor to the televisual Bad Man of today. Montana's impact, legacy and popularity continue to resonate in American popular culture and inspire countless quotations and references to iconic scenes ("Say hello to my little friend!") in an analogous manner to the popular Bad Men of today (a phenomenon that will be explored in depth later in this essay). Many of America's cinematic triumphs are built upon the compelling effect a villainous protagonist can have upon its audience—an effect that American television has recently learned to capitalize upon in this fandom-fertile age of media convergence. In summing up his historical analysis of American onscreen masculinity, Martin argues that "men alternately setting loose and struggling to cage their wildest nature has always been the great American story" (84).

Due to the mainstream acclaim, exposure and pervasive domestic presence of American quality television, many cult Bad Man television stars have been elevated to a status comparable to their Hollywood counterparts. The line separating the popular film and television has become increasingly blurred with many high profile actors taking on major roles across the two mediums. For this reason, I will be drawing on the work of Jackie Stacey in my approach to analyzing *Breaking Bad*'s audience identification with Walter White. In *Star Gazing: Hollywood Cinema & Female Spectatorship*, Stacey investigates the processes behind female spectators' identification with Hollywood stars in the 1940s and 1950s. Although these female Classical Hollywood stars and the contemporary Bad Man television characters may initially appear as polar opposites, I contend that the identification processes of their respective audiences may not be so dissimilar. The following section of this essay will apply Stacey's empirical research approach to the online accounts and paratexts generated by the fandom of *Breaking Bad* in order to investigate how the series' portrayal of masculinity has been received and reflected by its audience.

"Baddicts": Fan Tributes and Quotability

In its pilot episode, *Breaking Bad* introduces Walter White as an economically challenged chemistry teacher who is diagnosed with terminal lung cancer. Faced with this diagnosis, Walter resorts to using his extensive knowledge of chemistry to manufacture and sell methamphetamine, aiming to secure a financial future for his wife Skyler (Anna Gunn), teenage son Walter Jr. (RJ Mitte) and unborn daughter. More so than any of its Bad Man contemporaries, *Breaking Bad* deliberately inspired much debate amongst critics and audiences over whether its leading man is the "real" protagonist of the series and therefore the "correct" object of the audience's sympathy and support. This debate over identification was fought by two broadly defined opposing sides: viewers that supported Walter's transformation from a mild-mannered chemistry teacher to the ruthless drug lord "Heisenberg," self-identifying and labeling themselves as "Team Walt," and viewers that gradually came to condemn his increasingly immoral decisions and behavior. The degree of Team Walt's support for Walter varied between respective viewers, but the group can be collectively defined by their apathetic attitude towards the ethical issues raised by Walter's actions, or the outright dismissal that these issues even exist.

Conversely, the opposing group of viewers applied real-world ethics and engaged with the moral issues raised by the series, typically condemning Walter for his immoral decisions. The voice of this group emerged primarily in an appalled response to Team Walt's continued support of Walter. In an article for the *New Yorker*, television critic Emily Nussbaum candidly stated, "I will say it: some fans are watching wrong. This is the Bad Fan who didn't see it as abusive when Walt lied to Skyler nonstop; or when he sexually assaulted her in the kitchen; or when he overrode her restraining order and forced himself back into her home; or when he turned Walt, Jr., against her." This position against Team Walt was reiterated by feminist blogger A. Lynn, who bemoaned and condemned the way in which many viewers, in her view, had misinterpreted the intention of the series: "Listen, I love Breaking Bad and I love that Vince Gilligan has created a show where he 'explore[s] a world where actions do have consequences.' What I don't love—and what scares me—is that **people either didn't pick up on the fact that we're not supposed to be rooting for Walt anymore or they know and choose to root for him anyway**" (bold in original text). Such audience identification can be analyzed in relation to Stacey's conception of "cinematic identificatory fantasies," a term she uses to define the processes of identification that a spectator experiences while watching a Hollywood star. Stacey notes that these identificatory fantasies often manifest as feelings of "devotion, adoration and worship" for the star as well as an aspiration to become more like them, by transcending or "losing oneself" through the process of identification (138–151).

The cinematic identificatory fantasies of masculinity at work within *Breaking Bad*'s Team Walt fandom are perhaps most easily observed through the countless tribute videos to Walter White on YouTube. By compiling the highlights of Walter White's journey throughout the narrative of the series, these videos recreate, condense and amplify the processes of identification between Walter and the audience. Furthermore, every video features a comments section, encouraging viewers to express their own reactions, creating a fantastic range of unprompted audience accounts to draw conclusions from.

The video "Rise of Heisenberg: 10 Most Awesome Breaking Bad Moments" by the YouTube channel Geoff Freund represents a clear example of this breed of fan tribute video. The video is a compilation of scenes from the series, presented as a Top 10 countdown, with every clip prefaced by a title (e.g., #7 "I am the one who knocks"). Notably, all of the 10 "most awesome moments" involve scenes in which Walter displays his ambition and ruthlessness through acts of intimidation and/or violence. In doing so, the video effectively simplifies *Breaking Bad*'s original presentation of Walter's identity through the use of montage. While the series portrays Walter as an intricate, morally ambiguous and multi-faceted character, fan tributes such as the one in this video highlight the hyper-masculinity of Walter White while neglecting his less desirable attributes entirely (such as his cowardice, greed and narcissism), transforming the series' intended character portrayal into a simplified message of character propaganda. By prioritizing the presentation and reverence of masculinity over moral complexity, these videos sacrifice the identificatory depth of the series in favor of a basic connection to the viewer's favorite aspects of Walter's character.

Many YouTube users have voiced their admiration for Walter's Bad Man persona in the video comments, indicating a vicarious sense of excitement and pleasure experienced through Walter as a narcissistic conduit:

> minnie mathieu
> *4 days ago*
> I just watched the one where Walt hit the thugs with his car last night.... MIND BLOWN!!... can't believe I missed that episode when it aired.
>
> ALiBi212x
> *1 week ago*
> Bryan Cranston's Heisenberg voice is the most badass thing ever
>
> mgn1990
> *1 month ago*
> The "Say my name" scene puts a huge smile on my face. Every. Time. Breaking Bad is one of the best shows ever made.
>
> Loïc Retournard
> *4 days ago*
> "This... is not Meth!" BOOM
>
> Zeleste Neo
> *1 week ago*
> My all time favourite is definitely "Stay out of my territory": This is it were the badassness began! Oh yeah man!
>
> César Inzunza
> *4 weeks ago*
> How do you know you're insanely attached to this show? Because you can recall what these 10 scenes are about just by reading a word before watching them [all comments quoted directly from YouTube 09/01/14].

This final comment from César Inzunza explicitly points out what is hinted at by all the others: the quotability of these scenes is a vital component of both Walter's narcissistic appeal to audiences as well as, more broadly, the appeal of *Breaking Bad* as a successful, fandom-inducing piece of popular culture. Every scene chosen for the "Top 10 Most Awesome Breaking Bad Moments" compilation has been immortalized by a 'badass' one-liner. Examples of this include "Stay out of my territory," "I am the one who knocks," "Tread Lightly," and "Say my name," just to name a few. Although meaningless to the uninitiated, these simple quotes are enough to evoke specific meanings and memories of the series within *Breaking Bad*'s fan community. This phenomenon can be illuminated by Umberto Eco's 1985 essay "'Casablanca': Cult Movies and Intertextual College," where he describes the criteria necessary to create a postmodern cult movie:

What are the requirements for transforming a book or a movie into a cult object? The work must be loved, obviously, but this is not enough. It must provide a completely furnished world, so that its fans can quote characters and episodes as if they were part of the beliefs of a sect, a private world of their own, a world about which can play puzzle games and trivia contests, and whose adepts recognise each other through a common competence [3].

This enhanced "quotability" effect plays a major part in the formation of *Breaking Bad*'s vast and devoted fandom, and consequently its status as a seminal cult text. *Breaking Bad* has been so successful in creating its own "completely furnished world" that the series' fans (or "Baddicts" as they are sometimes called) are able to quote the series to each other as if it were a private cult language or code. More specifically, the construction and presentation of Walter's memorable quotes strengthens the character's narcissistic bond with the audience and subversively encourages fans to adopt, imitate and reappropriate these scenes outside of the world of the series. In addition to the fan-made compilation videos, the series' quotability can be observed in the use of visual memes. In the examples pictured below, three iconic Walter White quotes have been reappropriated in such a way that only fans familiar with the original text will appreciate the reference.

This recycling, reappropriation, and repetition of both visual memes and memorable quotes may be understood in the context of what Jackie Stacey labels as "extra-cinematic identificatory practices." Stacey defines these

"You're goddamn right." From the episode "Say My Name" (5.07). This popular meme is often used in the comments section of social media sites as a humorous means of expressing agreement. Image originally from imgur.com, September 21, 2014.

identificatory practices as behavior conducted by the spectator that demonstrates a desire to resemble, imitate or copy aspects of a star's identity (160–167). In relation to identifying with classical Hollywood stars, these practices would often involve attempting to recreate a popular actress's hairstyle, or acting out a well-known movie scene with friends. In the ongoing age of social media, examples of these practices can now be observed through the regular re-use of pop culture quotes as seen in the *Breaking Bad* examples above. The use of Walter White in memes and quotes demonstrates fans' devotion and admiration of the series and a specific performative mode of identification with the character Bryan Cranston portrays. By wearing

"I am the one who knocks." From the episode "Cornered" (4.06). This is one of many popular t-shirt designs featuring Walter and/or an iconic *Breaking Bad* quote. Image originally from redbubble.com, September 21, 2014.

a t-shirt featuring Walter White's face and a memorably intimidating quote, a fan is demonstrating their devotion, aspiration and adoration for the character, unconsciously or otherwise. Similarly, by using Walter's face in a meme designed to convey enthusiastic agreement ("You're goddamn right!"), fandom has transformed Walter into an online authority figure, as online fans adopt and imitate the hyper-masculine aspects of his persona to voice their approval and appreciation.

Again, these visual memes/quotations demonstrate fandom's ability to break down and simplify *Breaking Bad*'s complex themes into easily-digested symbols of masculine propaganda, perhaps to an even greater degree than the fan-made tribute videos. The manner in which these "badass" quotes are reduced, removed from context and then reappropriated through active fandom is symptomatic of the audience's strong identificatory fantasies with Walter's hyper-masculine Heisenberg persona. By selecting and simplifying the scenes and quotes that best represent fandom's favored representation of Walter, fans are able to act out their extra-cinematic identificatory practices

40 Hero or Villain?

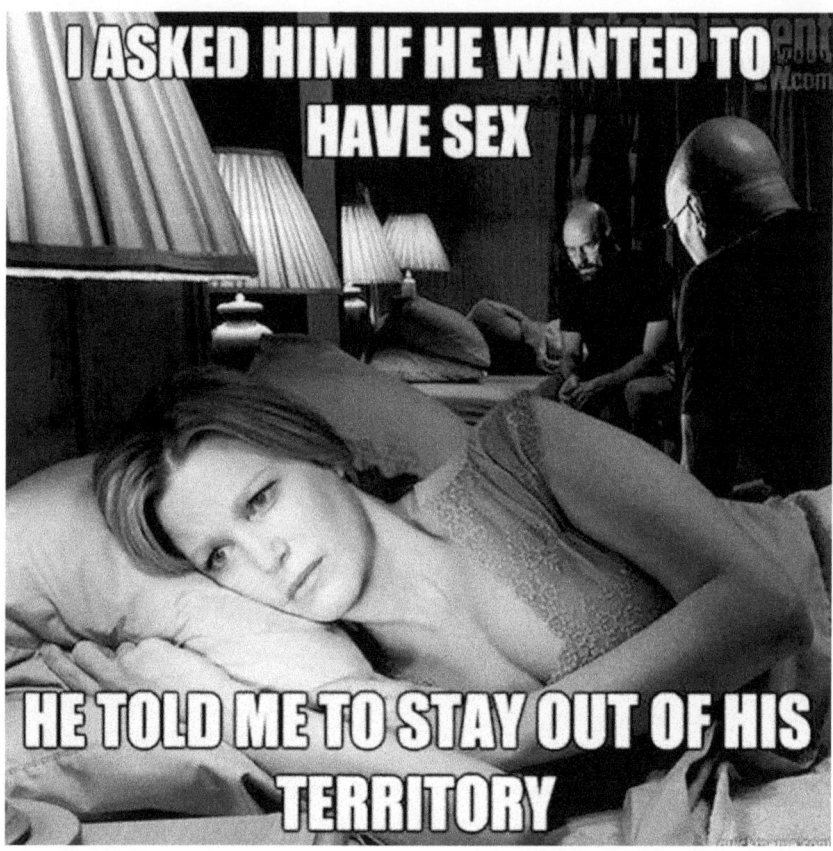

"Stay out of my territory." Parody of a scene from the episode "Over" (2.10). Image originally from badassmemes.com, September 21, 2014.

with ease, inadvertently constructing a legacy for Walter White which does not represent the full range of his in-show identity or the ironic manner in which his Heisenberg identity is presented.

Stacey's extra-cinematic identificatory practices can be observed today at their most extreme through the relatively recent rise of parody user accounts for fictional celebrities in social media. Walter White, along with many other popular television characters, now has several Twitter, Facebook and YouTube accounts in his name, which regularly post about his drug-related exploits and daily life. Here, devoted fans of *Breaking Bad* are taking this performative mode of identification into entirely new territory by actually adopting the identity of Walter in an online capacity. These fan-accounts often keep up the pretense that the character they portray is living a "real life" outside of the fictional reality of the series through regular Tweets.

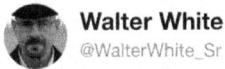 **Walter White**
@WalterWhite_Sr

Now you can buy the clothes I wore @ waltswardrobe.com Some good ideas for Halloween. Just don't come near my territory! #BreakingBad

12:49 AM - 19 Oct 2013

Tweet taken directly from Twitter, September 21, 2014.

The heightened presence of Stacey's cinematic identificatory fantasies and extra-cinematic identificatory practices is not limited to the contemporary audience's identification of Walter White. As a result of their prevalent presence and popularity, every successful Bad Man enjoys a similar level of admiration, imitation and media transcendence through the active identification of television fandom. Bad Men such as Tony Soprano, Dexter Morgan (*Dexter*), and Omar Little (*The Wire*) have all inspired an enduring cultural presence and have been fully embraced by their respective fandoms, transforming the characters into a mythical antiheroes in the realm of popular culture.

Graffiti of Tony Soprano in honor of the late James Gandolfini. Image originally from flickr.com, September 21, 2014.

42   Hero or Villain?

Another popular internet meme on social media sites in the same vein as Walter White's "You're Goddamn Right" meme featuring Dexter (Michael C. Hall). Image from quickmeme.com, September 21, 2014.

A t-shirt design featuring a popular quote ("Come at the king you best not miss") from popular *The Wire* character Omar Little (Michael K. Williams). Image from pinterest.com, September 21, 2014.

Hating Skyler White

Notably, fandom's positive reception and embrace of these male protagonists has not been extended to the female supporting characters of the Bad Man series. In fact, many female supporting characters have endured the opposite effect; entering the domain of popular culture as targets for fandom's hatred and active online abuse. As this essay will go on to argue, this overwhelming "hate-following" for supporting female characters is a direct result of the Bad Man series' ability to inspire strong audience identification with their leading protagonists as symbols of American masculinity. In December 2013, Kate Aurthur published an article on media/news website Buzzfeed titled "16 Female TV Characters Who Were Cyberbullied in 2013." The list includes Dana Brody (*Homeland*), Betty Francis (*Mad Men*), Debra Morgan (*Dexter*) and includes various memes and online comments as evidence of the harassment of these characters. Topping the list is *Breaking Bad*'s Skyler White.

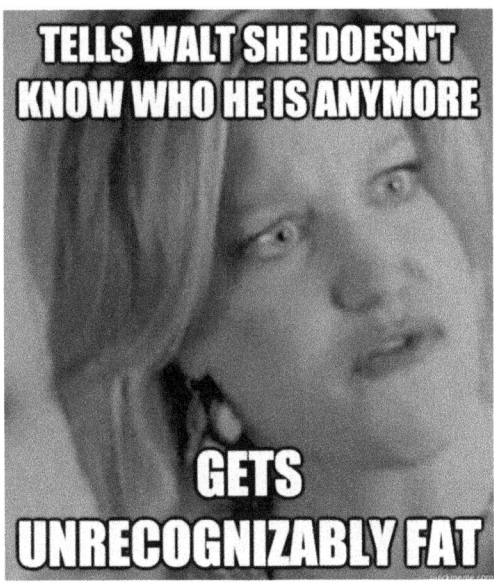

Image from Kate Aurthur's article, "16 Female Characters Who Were Cyberbullied in 2013," Buzzfeed, March 22, 2014.

Described by Vince Gilligan as an early "voice of morality on the show" (Brown), Skyler serves as one of Walter's biggest obstacles and opponents due to her early suspicions and moral outrage at her husband's activities. Despite her relatively ethical position and psychological victimization at the hands of Walter in later seasons, Skyler has generated a considerably larger

44   Hero or Villain?

hate following than her comparatively immoral husband. As of August 30, 2015, 30,111 people had "liked" the "Fuck Skyler White" (FSW) fan-made Facebook page which serves as a social space for fans to express their hatred for the character, usually alongside their respect for Walter. Examples of comments on the FSW page are listed below:

It's chicks like Skyler who make fellas like W.W. SNAP!!
22 February 2014 at 18:37

the actor and the character are both a bitch
7 October 2013 at 13:07

Wow you're seriously annoying. You wanted Walt to leave just because he has two phones… how suffocating.. I HATE YOU! >:C
21 June 2013 at 00:01

Just after watching episode 3 season 3, and all I can say is, Skyler you absolute twisted "scank" fucking whore, you definitely have to be related to my ex!
Skyler, you deserve the cancer! Walter FTW!
5 February 2013 at 11:10

You fucking suck! The only reason why people hate u so much is because you are the epitome of what women are capable of
22 April 2013 at 01:45

[All comments taken directly from the FSW Facebook page, August 7, 2014.]

In response to this online "Skyler hate" phenomenon, Anna Gunn, the actress who portrays Skyler, wrote an article for the *New York Times* arguing that her character has not been judged by the same moral standards as Walter. In it, Gunn wrote, "Could it be that they can't stand a woman who won't suffer silently or 'stand by her man'? That they despise her because she won't back down or give up? Or because she is, in fact, Walter's equal?" In response to Gunn's article, several members of the FSW page posted this internet meme.

Once again, this use of visual

Image from the FSW Facebook page, August 7, 2014.

memes demonstrates fandom's ability to simplify and reduce complex characters into woefully incomplete caricatures. The same group of fans who immortalized Walter's toughness, ingenuity and intimidating quotes, while conveniently disregarding his self-importance, greed and vanity, have applied the same selective means of representation to Skyler, only this time focusing on her negative traits and omitting her more sympathetic, humanizing features. In *Breaking Bad*, Skyler White is an emotionally, morally and psychologically complex character, who is frequently shown to be pragmatic, intelligent and sensitive but also occasionally controlling, emotionally fragile and corruptible. In certain corners of *Breaking Bad* fandom however, Skyler White is simply an annoying bitch who cheated on Walt with her boss and keeps getting in the way of Walt's "badassery."

This distorted perception may be understood in relation to Geraldine Harris' conceptualization of postmasculinity. In her article "A Return to Form? Postmasculinist Television Drama and Tragic Heroes in the Wake of The Sopranos," Harris collectively categorizes Bad Man series such as *The Wire*, *The Sopranos*, *Mad Men* and *Sons of Anarchy* as examples of "postmasculinist" television drama in the same way that shows such as *Ally McBeal* (1997–2002) and *Sex and the City* (1998–2004) are grouped under the rubric of postfeminism (443). Harris notes that, like postfeminism, these postmasculinist dramas display an "ambiguous/ambivalent relation" towards twentieth-century movements concerning the "de-centering" of the white masculine subject (443). Harris goes on to argue that these postmasculinist dramas create fictional worlds that are "dominated by misogyny, homophobia, and racism whilst (apparently) establishing 'ironic distance' from these attitudes" (443). According to Harris, the Bad Man series reinscribes and reclaims the antihero position for the white male and promotes masculinity as a "sign of powerful cultural capital" (459).

Harris also notes that these shows commonly portray a protagonist who is "in control, but on the verge of losing it to "emasculating forces." This is certainly true of Walter, who engages in a constant battle with Skyler for the domestic control of their household throughout the series. Harris argues that within these postmasculinist works, "women's relative rise to power does not seem to have been accompanied by a significant shift in focus" (460). As later seasons of *Breaking Bad* do attempt to shift the focus away from Walter while undermining the legitimacy of his power and control, fans who identify most strongly with Walter as a leading Bad Man character may view Skyler as a greater threat to Walter's masculinity (and vicariously their own), than any standard supporting character, thus inciting the increased level of hatred. It is worth noting that none of Walter's male adversaries, such as his federal agent brother-in-law Hank Schrader (Dean Norris) or rival drug lord Gustavo Fring (Giancarlo Esposito), have received such a venomous reaction from

Team Walt viewers despite the threat they pose to Walter's continued success. As Skyler does not possess, and actively undermines, the masculine qualities that define Walter and his adversaries, spectators who identify most strongly with the series' portrayal of masculinity will not extend their multiple points of identification to her character and may actively disregard and ridicule Skyler's own struggles, objectives and character development.

The extent to which hatred for Skyler can be directly related to real-life misogynist attitudes is open to debate. Towards the end of her article, Anna Gunn argues that most of the "Skyler-hate" is representative of these viewers' own perception of women and wives in society. She writes, "Because Skyler didn't conform to a comfortable ideal of the archetypical female, she had become a kind of Rorschach test for society, a measure of our attitudes toward gender" (Gunn). Vince Gilligan has voiced a similar sentiment, condemning the Skyler haters as "misogynists plain and simple. People are griping about Skyler White being too much of a killjoy to her meth cooking murdering husband? How could you have a problem with that?" (Brown).

Statistics provided by the administrator of the FSW Facebook page reveal that as of August 2015, 79 percent of its 30,111 fans, were male and 49 percent were aged between 18 and 24 years old (statistics taken 08/30/15). I immediately acknowledge the limitations of such raw data—without the corresponding demographics of *Breaking Bad*'s viewership as a whole, any conclusions drawn from these Facebook figures alone can be called into question. Nonetheless, on a superficial level at least, this predominantly male adolescent "hate following" does appear to support Gunn's assertions, as well as the application of Harris' postmasculinist theory to the Bad Man television series.

Despite the commendable artistic merit and intentions behind many of these Bad Man series, greater care should be taken with regards to the portrayal and reception of both the leading male and supporting female characters within them, from the perspective of television creators and fandom alike. Although the hyper-masculinity of Walter White's Heisenberg persona is arguably presented from a critically ironic perspective within *Breaking Bad*, irony itself is a double-edged sword, with the potential to either undercut or perpetuate the very attitude it aims to critique. At its worst, the close mirroring relationship between television and its audience is capable of endorsing, circulating and reinforcing negative social attitudes such as prejudice, racism and misogyny. But at its best, this connection is able to educate its mass audience, delicately contributing to the eradication of such destructive beliefs. The American Bad Man appears stronger than ever, but with careful repositioning and reconstruction of its audience's identification patterns, the inadvertent echoes of misogyny that surround the Bad Man series may soon be regarded as regrettable positions of the past.

WORKS CITED

Aurthur, Kate. "16 Female TV Characters Who Were Cyberbullied in 2013." *Buzzfeed*. 12 Dec. 2013. Web. 22 Mar. 2014.
Brown, Lane. "In Conversation: Vince Gilligan on the End of Breaking Bad." *Vulture*. 13 May. 2013. Web. 22 Mar. 2014.
CD13Production. "Breaking Bad Tribute—Choices That I've Made." Online video clip. *YouTube*. YouTube, 15 Oct. 2013. Web. 1 Sept. 2014.
Eco, Umberto. "'Casablanca': Cult Movies and Intertextual Collage." *SubStance* (1985): 3–12. Web. 18 June 2014.
Freund, Geoff. "Rise of Heisenberg: 10 Most Awesome Breaking Bad Moments." Online video clip. *YouTube*. YouTube, 2 Mar. 2014. Web. 1 Sept. 2014.
"Fuck Skyler White" (Community Page). Facebook. Founded 14 Aug. 2011. Web.
Gunn, Anna. "I Have a Character Issue." *New York Times*. 23 Aug. 2013. Web. 14 July 2014.
Harris, Geraldine. "A Return to Form? Postmasculinist Television Drama and Tragic Heroes in the Wake of The Sopranos." *New Review of Film and Television Studies* 10.4 (2012): 443–463. *Taylor & Francis Online*. Web. 12 May. 2014.
Lynn, A. "Team Walt? Really?" *Nerdy Feminist*. 16 Sept. 2013. Web. 20 Aug. 2014.
Martin, Brett. *Difficult Men*. New York: Penguin Press, 2013. Print.
Nussbaum, Emily. "That Mind-Bending Phone Call on Last Night's 'Breaking Bad.'" *The New Yorker*. 16 Sept. 2013. Web. 22 Mar. 2014.
Stacey, Jackie. "Feminine Fascinations." *Star Gazing*. London: Routledge, 1994. 126–175. Print.

Breaking It Down
Expressing Black Selfhood and Subverting Binaries of Good and Bad in Key & Peele

HILARIE ASHTON

Dig, if U will, this picture: Barack Obama sits in an office, in front of a camera, looking recognizably serious and poised. He's ready to address an attentive nation, and yet he isn't alone. Looming behind his chair is a tall, bald man in a suit with rings on each lanky finger. The President begins to speak, graveling through a slight stammer, and when he pauses, the bald man growls in taciturn response. If you keep watching the comedy sketch (for that is what it is), the tall man's energy will crescendo. Though Obama will stay seated, Luther, who we will learn is his Anger Translator, will stalk the performance area. *Key & Peele*, the (pop-)critically lauded sketch comedy series that aired on Comedy Central from 2012 to 2015, launched this inextricable pair. With them, their creators Keegan-Michael Key and Jordan Peele have introduced the idea of an emotional translator into our pop-cultural consciousness. Indeed, Luther's sketches have made such an impact that President Obama himself helped bring down the fourth wall by performing with Luther (via Key) at the 2015 White House Correspondents' Dinner. In this essay, I argue that Key and Peele thoroughly shake up a version of a "good guy"/"bad guy" dichotomy through the pairing of a personage everyone knows (Obama, played by Peele) with an imagining of a certain kind of counterpart (Luther, played by Key); each character's roles and positions, but especially Luther's, put them in a liminal space where good and bad are less easily definable.

The unsettling of a common binary is in part a function of (sketch) comedy itself—what Zadie Smith, in her *New Yorker* February 25 profile of the duo, called "the antic freedom inherent in sketch, which, unlike sitcom, can present many different worlds simultaneously." This ability seems even truer of sketches that center on just two main actors: Key and Peele each play many roles throughout the series, trading off taking "good" and "bad" positions in different sketches (and, of course, sometimes dispensing with the binary entirely, as I now will do with scare quotes, considering them implied). They trouble this divide to such a degree that the terms good and bad often lose their traditional meaning: in the Luther sketches, as I'll suggest, honesty versus expectation, impulse versus inhibition, freedom versus duty, and said versus unsaid are some of the valences that the emotional translation conceit shakes up. As a recurring character, Luther is a watershed in the subversion of a good guy/bad guy paradigm, I argue. While letting his inhibitions go might seem bad, it's also good in the terms of the emotional honesty the sketch sets up. At times, the mutable slash between bad and good nearly disappears, since Luther functions as good in relation to the audience by virtue of telling them one complex and relatable truth, and yet the version of truth Obama is telling is both performatively structured and required of his position, as well as highly racialized, as I'll show.

To this end, I examine several of *Key & Peele*'s Luther sketches to flesh out the show's transformation of a good guy/bad guy division. These close readings suggest a conception of Key and Peele, shored up by Zadie Smith, Shari Stenberg, and other critics, as more than just comedians: they are also advocates, through their comedy, for a less racist, less sexist, less homophobic world. Good and bad, *Key & Peele* shows us, are just as much structural concepts as individual ones, and in that sense, they frame Luther's narrative. In this way, Key and Peele the comedians are useful challengers to a frighteningly prevalent notion of American society as "post-racial," one that became unavoidable in media and popular discourse after Obama's election. Key and Peele themselves trouble this term implicitly and explicitly throughout the run of the show, both within sketches and anytime they use the stand-up intro to discuss their biracial heritages, for example. They also take on and recreate different moments in Black history from slavery to Martin Luther King and beyond. In season two, they recreate a New Black Panther Party whose biggest challenge, Key's chairman character announces at a press conference, is to exist in this "era of so-called racial harmony" and Black death at the hands of police, with Peele popping out from behind Key and inserting his face into the shot at ridiculous angles ("Dueling Hats"). The sketch, like so many of their others, melds a serious vein of social commentary with slapstick humor.

With the advent of the Black Lives Matter movement and the increase in anti–Black violence that necessitated it, this post-racial space that is so often under *Key & Peele*'s analysis becomes more and more imaginary. Very recently, in 2016, the virulent and public expressions of racism and bigotry by Republican then–presidential candidates Donald Trump and Ted Cruz has spotlit the kind of angry whiteness within an already racist system that most progressive white people have historically had trouble seeing clearly enough. Many journalists and academics have written incisively about this dynamic; Mychal Denzel Smith, for one, wrote in *The Nation* in February 2016 that, because of its appeals to anti–Black, anti–Mexican, and anti–Muslim bigotry, "Trump's candidacy has made explicit appeals to racism politically viable again."

Meeting Luther

Luther's first appearance on *Key & Peele* was in the show's pilot, which aired on January 31, 2012; he made regular appearances in sketches up through the last episode of the show in 2015. His introduction opens much in the way I described in my opening paragraph. Peele serves up a spot-on Obama impression, mapping his hand motions, stutters, pauses, and turns of phrase, and provides a telling (and tongue-in-cheek) introduction on which the sketch and those that follow will turn: "Now, before I begin, I just want to say that I know a lot of people out there seem to think that I don't get angry. That's just not true. I get angry a lot. It's just that the way I express passion is different from most. So just so there's no more confusion, we've hired Luther here to be my anger translator." The camera then pans up diagonally from Obama to reveal Luther looming in the background of the shot; his first word is a growled "Hi." He uses the camera much more than Peele does: he comes to the center and ranges forward so that he's filling the borders of the screen. He also runs around the space between Obama and the fireplace, his face barely visible, letting his voice and body do the work of drawing the viewer's attention, at one point jumping up and gluing his face to the camera.

Luther's anger in this scene works in a visual and embodied way, uncoiling and rising the further the scene goes. It explodes, brilliantly, in a rant that is echoed, like Obama's introduction, as part of the pattern of each Luther sketch. This first one is rich and layered in linguistic and cultural reference: "I have a birth certificate. I have a birth certificate. I have a hot diggity, daggity, mamase mamasa mamakusa birth certificate, you dumb-ass crackers!" With these lines, Luther's language and tone open out into Black vernacular (the last words echoing the refrain of Michael Jackson's *Thriller* cut "Wanna Be Startin' Somethin'" (1982), which is in turn taken from Manu Dibango's 1972 "Soul Makossa"). This doubled cultural quote is particularly interesting when considered in light of Obama's Kenyan birthplace; its primary function, though, seems to be to add to the verbal code-switching that Luther's presence allows Obama to participate in. That is, the President's formal (coded White) language gives way in Luther's translation to a closer approximation of African American Vernacular English (AAVE); in the conceit of the show, this is where Obama's truest emotions live. In this way, the first sketch of the series sets the tone for much of the linguistic and cultural (and embodied) work that the biracial duo will take on in the series to shine light on and tear down racial stereotypes, and also to limn the shifting boundary between good and bad.

In her unfortunately titled but sharply observed "Color Commentary," a 2013 review, *New Yorker* television critic Emily Nussbaum recounts seeing a Luther skit before any other part of the show, describing it as feeling "kind of like a one-joke gag." She remarks later that in rushing to categorize the show, she had

> committed the dumbest possible error, slotting it into a ridiculous niche: the comedy show with two light-skinned black guys. No one calls Jimmy Fallon's program "the white-boy show." No one describes the *Late Show with David Letterman* as "the seems-like-he's-about-to-snap-Caucasian-codger show." Yet I'd not only wedged *Key & Peele's* comedy in with other Afrocentric sketch shows, from *In Living Color* to *The Chappelle Show* [sic]; I'd reduced the pair's biracialism [sic] to a kind of gimmick.

Nussbaum's openness here plays nicely with the kind of good/bad transcendence that I'm arguing that Key and Peele demonstrate; importantly, though, it also sets up her awareness of her own subjectivity compared to theirs. The show does not exclude white audiences, but it refuses to accept the default of whiteness, and part of what it teaches white viewers is to enact that same refusal. Race is not always the show's central issue, as she realizes, but it's part of the fabric of its reality in a flexible and subtle way.

Nussbaum continues usefully, framing Key and Peele's contributions in terms of the transformations they enact in most of the series' sketches, while continuing to place them in a strong comedic and creative lineage: "Like many of the best, most transgressive comics, they treat human behavior as a form of drag, shape-shifting with aggressive fluidity.... Key and Peele's true passion is intra-black-male drag." Part of their shape-shifting is verbal, of course: not only do they move seamlessly between different varieties of English, but they also point directly to that codeswitching in several ways. Luther is just one example of the importance of codeswitching to the show; in a later sketch, a substitute teacher serves as his analogue by bringing his self-identified inner city teaching experience into a largely White classroom, taking roll while steadfastly claiming that each student is mispronouncing their own name ("Substitute Teacher"). He introduces himself memorably: "All right, listen up, y'all. I'm your substitute teacher, Mr. Garvey, I taught school for 20 years in the inner city, so don't even think about messing with me. You all feel me? Okay, let's take the roll here. Jay-quellin, where's Jay-quellin at? No Jay-quellin here?"

When the student named Jacqueline corrects him, he pauses, exasperated. To the next student, "I'm for real. I'm for real, so you better check yourself.

Dee-*nice*, is there a Dee-*nice*? If one of y'all say some silly-ass name, this whole class is going to feel my wrath." By the end of the sketch, he's exploded along Lutherian lines, breaking his clipboard over his knee, jumping up and down, and sweeping things off the table. The punchline intensifies with Peele's role as the class' only Black student answers Mr. Garvey's "Tim-*mothy*" with "*Pree*-sent." Peele's character is the only student who shifts in the direction Mr. Garvey is imposing. Implicit in the sketch is the awareness that it is more often a white teacher imposing on Black students than the reverse, and so in this sense, the division between good and bad paradigm is destabilized even further. When the sketch is revisited in season three, the dynamic has shifted: the students accept Mr. Garvey's (mis)readings of their names while he continues to question the default of whiteness ("Les Mis"), another instance of pointing out the quiet, racialized danger of the status quo even as the system conditions students to think that it is good.

In these sketches (and, presumably, in reality as President), Obama has to play a "good" role and say more of what's expected than (perhaps) what he really thinks. In Zadie Smith's 2015 *New Yorker* profile of Key and Peele the comedians, she elaborates this characterization with one of the fourth wall-bending facts of the show: "Later that year [2012], Obama asked to meet with Key and Peele, and wryly acknowledged the same desire. 'I need Luther,' he told them. 'We'll have to wait till second term.'" In their intro stand-up segment on the first episode of the second season, Key and Peele bend the fourth wall again, talking about having met the President in perhaps the same exchange that Smith quotes. Peele quotes him as having said, "By the way—don't let me do anything too crazy" ("Obama College Years").

I read these two comments as a partial admission by Obama himself of the kind of "good" and "bad" expectations placed on him as the first Black president: there are things he wanted to say that were not, at that point in his tenure, sayable. The kind of anger a white president might be allowed to express was off limits to him. Peele's portrayal of Obama is a very purposeful foil for Key's Luther, after all: the audience has to be able to believe that Peele *is* the President in order for the more outsize sides of Luther's comedy to work. We almost have to look past Obama to fully appreciate Luther, and it's that looking past, while recognizing Peele's Obama as familiar and believable, that allows the different levels of transcendence to occur.

Twice More with Feelings

In addition to the show, Luther sometimes turned up in short Web exclusives, like an unsourced video Comedy Central posted to YouTube in 2012. Titled "Obama in Therapy," the sketch shows both men sitting, Obama leaning back and Luther angled forward in a posture of tense muscle control. Obama is meeting with an off-screen therapist in this scene, and his summation of his relationship with Michelle Obama as "couldn't be better" is punctured by Luther's first line: "That bitch is crazy." When the therapist responds, "I need you to express your own emotion," Luther stands and stage-whispers, "Why you're telling him to get angry, dawg, that's my job," underlining his role in an economy of, as I said at the beginning, if not always strictly good and bad, then definitely socially acceptable and non. When the therapist transitions into asking Obama to express his anger on the pillow, Luther stabs it with a penknife.

His sometimes spastic and unpredictable movements aside, the primary way in which Luther embodies a version of goodness is by doing his job: translating Obama's anodyne statements into their real meaning, according to the translation conceit of the sketch. Luther's translation is full-bodied, marking a parallel to American Sign Language (ASL) interpretation. Here, Obama and Luther are speaking different languages with different physical and linguistic grammars, just as an ASL interpreter does when using English to communicate the signs of a Deaf person. So much of the grammar of ASL is shaped by embodied space and movement. Present and past are indicated with respect to the body, for instance, imbricating time with space. ASL is too often incorrectly cast as a "bad" or pidgin translation of Standard American English, when it actually uses its own grammar and is more closely tied to French Sign Language than to spoken English. In many ways, Luther's translation is more akin to interpretation anyway, since physical motion and use of space is such an important part of it, broadening still the valences of modal translation to include motion as communication and boundary crossing, as well.

Ultimately, and especially after appearing in several sketches, Luther's unpredictability itself becomes part of his goodness—he is, after all, being as honest as his job allows him to be. This reading of his metronomic loss of control allows for the character's complexity in a way that figuring it as simply bad does not. When he succumbs to the heat of the moment—of actual feeling, sometimes Obama's and sometimes his—he rejects the frame, and reveals himself to be more of a real person. Alongside that, the sketches usually include a corrective to Luther's explosive release of feeling, one that gets him back on the path of duty and expectation. In their first appearance, for instance, when Obama gently reprimands him for getting out of control, Luther neatly inverts Obama's own hand gesture, pointing at his own face while chiding himself, "Dial it back, Luther, daaaamn" ("Series Premiere"). This moment shows both the comedians' facility with internal references as well as Luther's own awareness of the importance of his job. (The crux of one of the third season's sketches, in which Michelle Obama's anger translator,

Katendra, is introduced during a domestic spat between the Obamas, is when the two translators act out suppressed attraction while doing their jobs) ("Obama Shutdown").

In *Ugly Feelings* (2005), Sianne Ngai traces the politics and presence of emotions of blockage, like paranoia and envy, that sit in contrast to what she classifies as easily expressible emotions like anger. While it doesn't apply completely to the kinds of open emotions that Luther is translating, the idea of not having action is useful for interpreting Obama's perspective. In her chapter on animatedness, Ngai argues that "as we press harder on the affective meanings of animatedness, we shall see how the seemingly neutral state of 'being moved' becomes twisted into the image of the overemotional racialized subject" (91). Luther's version of motion, physicalized, seems to me to be much more within his own control—he does not *have* to move the way he does, but the sketches seem to imply that he could not translate Obama's emotions (of anger, but also of ugly feeling-adjacent emotions like frustration and bewilderment) as effectively without it. It's worth noting, too, that the passages Ngai analyzes in this section involve white authors problematically assigning motives to people of color, rather than two biracial men creating and expressing (mostly) biracial characters in a comedy series.

Scholarly Glances

Part of the reason I wrote this essay was to add to the woefully inadequate number of academic works on *Key & Peele*. The names "Keegan-Michael Key" and "Jordan Peele" appear twice in the JSTOR database's scholarly works (as of July 2016); I discuss those pieces, by David Gillota and Shari Stenberg, below. By contrast, the name "Dave Chappelle" (to whom the duo is often and not entirely fairly compared) appears 56 times. EBSCOHost has a similar ratio, capturing *Key & Peele* once and Dave Chappelle 32 times. Not every article that mentions him takes Chappelle on fully, of course, but he's part of the backdrop even if he's not the foreground.

In "Black Nerds: New Directions in African American Humor," Gillota centers Donald Glover and Key and Peele "at the forefront of a new trend in contemporary African American humor [because they] directly discuss race but do so in ways that reject the traditional signifiers of black popular culture and eschew the conventions of mainstream African American humor that have become commonplace since the early 1970s" (18). In the conclusion to his book *Ethnic Humor in Multiethnic America*, also published in 2013, Gillota remarks, "Ethnic humor in the future will likely be much the way it is in the present: contradictory, complex, challenging, and often fraught with anxiety" (154–155). Such an analysis seems right, and is only strengthened by the continued success of starring comedians of color who take on serious issues, among them Aziz Ansari on Netflix's *Master of None*, Tracee Ellis Ross and Anthony Anderson on *black-ish*, and Jerrod Carmichael on *The Carmichael Show*. Even shows dominated (or dictated) by white actors and viewpoints, like Lena Dunham's *GIRLS* and Tina Fey's *Unbreakable Kimmy Schmidt*, are being increasingly and rightly critiqued for the gaps in the worlds they depict.

When Gillota continues in his conclusion, gaps within his own analysis start to show: "The most sophisticated humorists will follow in the paths carved by comics like [Russell] Peters, Dave Chappelle, Sacha Baron Cohen, and Trey Parker and Matt Stone. These humorists will be willing to look beyond the rigid boundaries of ethnic identification and use humor to explore ethnic identity in more complex ways" (154–155). I'm profoundly surprised that despite his vanguard list, he doesn't mention any *female* humorists of color. He mentions Whoopi Goldberg only twice, once in the context of another scholar's work on comedians who've voiced animated characters, for instance, and he doesn't take on Mo'Nique or Retta, glancing uncomfortably and unfairly slightly across the spectrum of Black female comedic work. (He does consider Margaret Cho, but even his consideration of white female comedians is limited—he writes about Sarah Silverman at length only in his chapter about Jewish comedians.) While the popular imagination certainly tends to leave more room for male comedians, and then for white female comics over female comics of color, Gillota's book could have usefully engaged and helped to undo that very racialized and gendered lacuna.

The difficulty of talking both broadly and closely enough about artists without also mentioning (or not mentioning) their close counterparts is underlined by the unavoidable presence of Dave Chappelle in most analyses of *Key & Peele*; he is the progenitor that pop critics most often cite when thinking about their comedic genealogy. It's an uncomfortable relationship, in part because of comments Chappelle himself has made about the duo, telling *Esquire* in 2015 that "I'm happy that I can sit at home on a Tuesday night and watch *Key & Peele* do my show, and it doesn't hurt me," and also because the comparison isn't entirely fair. Though Key and Peele, like Chappelle, are men of color making a sketch comedy show named after them that takes on race and class, those points of convergence don't add up to a recreation of his work.

As Kenneth Ladenburg notes in his article on the racial politics of *It's Always Sunny in Philadelphia*, "humor is a powerful rhetorical device" (859). Of the show's pilot episode, "The Gang Gets Racist," he argues that "the hidden transcript of whiteness is made visible through the cast's humorous treatment of the divide between political correctness and racial naiveté" (860).[1] This idea of a hidden transcript of racial codes is laid openly bare in *Key & Peele*, less sardonically than in *Sunny*, but just as powerfully (and arguably more so, since *Sunny* rarely features any actors of color). In his introduction to his dissertation *Staging Alterity: The Ethics of Performing Difference(s)*, Joshua Abrams writes that the artistic works he examines in the context of a Levinasian ethics "begin to allow us to imagine a new universal subject, one who is defined in substitution, in the prostration of the self at the foot of the Other; this is a subject defined theatrically in obligation to another who appears before her/him" (5). The liminal and race/gender/class-conscious characters in *Key & Peele*'s leftist, multicultural world seem like they inhabit a similar ethical, outward-facing space.

While the edge to the Luther sketches' humor comes mostly from the stereotypical perception that President Obama, as a Black man, should not show anger, Luther's emotional release has another, slyer effect: it subverts the perception that professional, public figures can never show emotion (and that women are overemotional) by literalizing a male overemoter. Shari Stenberg's project in *Repurposing Composition: Feminist Interventions for a Neoliberal Age* (2015) is explicitly grounded in feminism, composition and rhetoric, and affect theory. She specifically characterizes the Luther sketches as "offer[ing] another perspective on Obama's allowable emotional range, providing commentary on which emotions an African American man may or may not display" (51). She also nicely sums up the debate over the racialized control over Obama's language, noting that "*Key and Peele*'s Luther skits inevitably lead to Obama's discipline of Luther—telling him to 'rope it in' or 'dial it back'—showing that Obama can't risk demonstrating outrage about racial discrimination" (52–53).

As I mentioned at the outset of this essay, *Key & Peele*'s own feminism is strongly threaded throughout the show's whole run. Until the sketch that gave Hillary Clinton her own translator, the trope had been the necessary—to a structurally racist, misunderstanding society, that is—subsuming of Black grammar (and meaning) into a mode of discourse palatable to white people. The appearance of Clinton's own anger translator in the fifth season episode "Y'all Ready for This" shifted the act of translation from unique to mundane—now, it appeared that everyone in office in the *Key & Peele* universe had one. Unfortunately, the sketch didn't open out the trope beyond that fact—even in the sketches where Michelle Obama got her own translator, the main axis was men translating for men and women translating for women. It would have been interesting to see that shaken up, which would have supported the show's feminism in a different way. In other ways, the show frequently riffed on the empowerment of women and on a liberated female sexuality (as in the sketches "MC Mom," "Y'all Ready for This," and "Airplane Showdown," all in the show's fifth season). "Y'all Ready for This," for instance, includes a sketch with male pirates singing a song about feminism (sample lyric: "We say 'yo ho,' but we don't say 'ho,' 'cause 'ho' is disrespectful, yo") and the sketch's dramatic reveal is that the pirate king is also a woman, played by Rebecca Romijn. This fluidity with roles, and riffing on the viewer's expectation, rounds out a transcendence of another good/bad dichotomy.

Luther and the Fourth Wall

Peele's Obama usually directs his remarks to the audience, a fourth-wall breaking device that's made possible by the sketches' typical frame of an address to the nation; it's part of what gives the framework of the sketch its realism. While Luther also speaks to the camera, he doesn't break the wall fully when he stays within the realm of translation. In a sketch that aired immediately after Obama and Romney's second debate in 2012, Luther translates Obama's "a time for reflection" as "I got my swagger back, bitches"; Luther here breaks the fourth wall in a way by pointing to Obama and saying, "This motherfucker here don't even need me. Strange, you know I'm Black right?" ("Bone Thugs & Homeless"). About a third of the way through the sketch, the fourth wall cracks further when Luther, referencing how Obama had shown Romney his "angry eyes" during the debate, points at Peele's (deadpan) face.

Perhaps the location in which it is most interesting to examine Luther as a liminal figure at the messy convergence of good and bad is Key's appearance with President Obama himself at the 2015 White House Correspondents' Dinner. Alongside the reference to needing Luther that Obama made in 2012, this is another moment of the bend in the fourth wall between the show and reality—in fact, with Key and Obama on stage together, the wall briefly disappears. Obama, perhaps the most adept comic talent the White House has seen, starts the bit on his own, remarking, "You know, I often joke about tensions between me and the press, but honestly, what they say doesn't bother me. I understand we've got an adversarial system. I'm a mellow sort of guy. And that's why I invited Luther, my anger translator, to join me here tonight." Key sidles in, yelling, "Hold on to your lily white butts!"

Obama later lists the slate of problems related to global warming, his emotions amping up as he goes. The exchange that follows reverses the roles the show's sketches have set up:

> LUTHER: Okay, I think they got it, bro.
> OBAMA: It's crazy! What about our kids? What kind of stupid, short-sighted, irresponsible, bull—
> LUTHER: Wow, hey!
> OBAMA: What?
> LUTHER: All due respect, sir, you don't need an anger translator. You need counseling.

In a reversal of the show's usual conceit, Luther is suddenly allowed to be the voice of proper comportment here, and Obama gets to let his feelings out.

In a world that frames expectations for people of color differently than it does for white people, Key's Luther and Peele's President Obama show us both the social pressure to live up to expectations and the socially conscious need to transcend them or to move around within them. As Peele remarked on NPR's *Fresh Air with Terry Gross* in 2013, "the reason both of us became actors is because we did a fair amount of code switching growing up, and still do." (Interestingly, the name given to the web version of the interview, "For Key and Peele, Biracial Roots Bestow Special Comedic 'Power,'" is odd, conjuring images of magic that veer toward the racially essentialist.) Ultimately, the kind of liminality that the Luther sketches reflect—the boundary straddling, et cetera—is central to the values of the show itself. As actors, writers, and creators, Key and Peele consistently break down societal stereotypes, sometimes even ending their sketches on an O. Henry-like twist. What Luther nearly achieves, as Zadie Smith intimated, is to say the unsayable. He is complexly drawn in interesting, challenging ways that make him an intriguingly disruptive figure to a good guy/bad guy paradigm.

Notes

1. The phrase "the hidden transcript of whiteness" appears to have been first used in *Staging Alterity: The Ethics of Performing Difference(s)*, a 2008 dissertation by Joshua Abrams that is available on Google Books. Abrams deploys it on page 123 to illuminate the work of artist Guillermo Gómez-Peña and how he "performs his Mexican identity through the cultural visions of the colonialist whites" (122).

Works Cited

Abrams, Joshua. *Staging Alterity: The Ethics of Performing Difference(s)*. Diss. CUNY Graduate Center, 2008. New York: UMI, 2008. Print.
"Airplane Showdown." *Key & Peele*. Comedy Central. 15 July 2015. Television.
"Bone Thugs & Homeless." *Key & Peele*. Comedy Central. 24 Oct. 2012. Television.
"Dueling Hats." *Key & Peele*. Comedy Central. 28 Nov. 2012. Television.
"For Key and Peele, Biracial Roots Bestow Special Comedic 'Power.'" *Fresh Air with Terry Gross*. National Public Radio. 20 Nov. 2013. Radio.
Gillota, David. "Black Nerds: New Directions in African American Humor." *Studies in American Humor* 28.3 (2013): 17–30. Accessed 30 July 2016. Web.
———. *Ethnic Humor in Multiethnic America*. New Brunswick: Rutgers University Press, 2013. Print.
Klein, Ezra. "The Joke Was That Obama Wasn't Joking." Vox.com. 27 April 2015. Accessed 30 July 2016. Web.
"Les Mis." *Key & Peele*. Comedy Central. 18 Sept. 2013. Television.
"MC Mom." *Key & Peele*. Comedy Central. 19 Aug. 2015. Television.
Ngai, Sianne. *Ugly Feelings*. Cambridge: Harvard University Press, 2005. Print.
Nussbaum, Emily. "Color Commentary." *New Yorker*, 30 Sept. 2013. Accessed 30 July 2016. Web.
"Obama in Therapy." *Key & Peele*. Comedy Central. 30 Jan. 2012. Television.
"Obama Shutdown." *Key & Peele*. Comedy Central. 16 Oct. 2013. Television.
Patches, Matt. "Dave Chappelle Speaks Out About His Comeback and Comedic Activism." *Esquire*, 20 July 2015. Accessed 30 July 2016. Web.

"Series Premiere." *Key & Peele*. Comedy Central. 31 Jan. 2012. Television.
Smith, Mychal Denzel. "Trump's Racism Didn't Scare Me. Now It Does." *The Nation*, 12 Feb. 2016. Accessed 30 July 2016. Web.
Smith, Zadie. "Brother from Another Mother." *New Yorker*, 23 Feb. 2015. Accessed 30 July 2016. Web.
Stenberg, Shari. *Repurposing Composition: Feminist Interventions for a Neoliberal Age*. Logan: Utah State University Press, 2015. Print.
"Y'all Ready for This?" *Key & Peele*. Comedy Central. 8 July 2015. Television.

Talk Bluntly and Carry a Pointy Stick
Violence and Verbal Complexity in Buffy the Vampire Slayer

TIRZA I. LEADER *and* DARCY MULLEN

Buffy the Vampire Slayer is a popular television show that brings youth and violence together in a regular, frequent, and unique manner. The premise of the show is that Buffy, a young, blonde high school (and later, college) student steps out of her stereotypically adolescent and feminine boundaries to combat forces of evil in her eponymic role as a slayer of "vampyres." Her innocence is tested in a series of encounters with bad guys—both bad guys as villains, and bad dudes that contribute to a complex composite of Buffy's sexuality. She is a protagonist that has added to the collective understanding of today's archetypical good guy—or good gal. For the seventh and final season *Buffy the Vampire Slayer* consistently held a Nielson Rating of 3.0 or higher, which means approximately 3,201,000 people sat down to watch every Tuesday night ("Neilson Ratings"). In 2013, TV Guide named it in the top 60 television shows of all time (Rousch 16–17). Many of the viewers of *Buffy the Vampire Slayer* lauded the show for its portrayal of strong female characters. The show, however, has also seen its share of criticism for the violence the heroine perpetrates. While critical work examines language, violence and masculinity in the show (such as Malgorzata Gosia Drewniok's 2012 "'You're Strong. I'm Stronger': Vampires, Masculinity and Language in *Buffy*") the relationship between rhetoric and violence has not yet been fully examined.

The show's unique look at youth and violence has come to the attention of several scholars.[1] A great deal of early research is concerned with the parallels between teen angst and supernatural violence. Since then, the scholarship has expanded to include academic niches and sub-niches that continue to captivate viewers, fan-fiction writers, and scholars alike. While the series is in a fantasy-realm, as Kathleen McDonald writes in "The More Things Change: Buffy and Angel Enact a Modern-Day Sentimental Novel," the show's "fantastic creatures and world of magic [are in] the midst of a reality that the average viewer would find quite within his/her normal societal expectations" (114). Highly relatable, the fantastical nature of the good guys/gals and bad guys do not distract from the reflection of life that the show provided viewers. As the show progressed and Buffy left high school, entered university, and subsequently withdrew from university, teen-angst grew into adult anxiety and anguish while the heroine still grappled with supernatural violence. *Buffy* continues to provide a unique opportunity to study the rhetorical consequences of violence.

The following analysis examines *Buffy the Vampire Slayer* in an attempt to understand a rhetorical or social connection between violence and verbal complexity for protagonists and antagonists—or good gals/guys and bad guys. While engaging contemporary rhetorical scholarship on *Buffy the Vampire Slayer*, we discuss the effects of violence and media on youth. Next, this essay explores the link between violence and verbal complexity. Following that, we present the data found in a statistical analysis of the relationship between the violence and the verbal complexity (or ability to communicate at a higher-level) of the heroine of *Buffy the Vampire Slayer*. Finally, a discussion considers these results and possible avenues for further study.

Violence and Media

Children today live with many images of violence in real world events. Though events in the greater part of society (e.g., September 11th, 2001, the war in Iraq, mass-shootings) instill fear in children for their randomness and scope it is the local violence in schools that causes the most damage due to its physical and emotional proximity (Edwards C01; Curry 5958). As Christine Jarvis writes in her 2001 essay, "School Is Hell," it is not surprising that schools are frequent settings for horror narratives in the popular media.

Media violence has often been cited as one of the primary reasons for the growing level of violence in American society (e.g., Savage 99–128). Much of this research can be traced back to Bandura's demonstration that children will model violent acts that they have experienced both directly and vicariously (1977, 1986). Notably, in the famous Bobo doll studies, children modeled an adult's violent behavior towards a Bobo doll after watching the adult's actions on a television (Bandura, Ross and Ross 3–11). Later researchers have expanded on Bandura's original ideas of modeling to show that media violence instigates violent behavior in unprovoked men and women (Huesmann and Miller 125–139; Paik and Comstock 516–546; Wood, Wong and Chachere 145–165; Zillman and Weaver 371–383).

Despite concerns about the relationship between media violence and aggression, media violence has not reduced (Pearl, Bouthilet and Lazar). In 2000, the American Academy of Pediatrics, the American Medical Association, the American Academy of Child and Adolescent Psychiatry, the American Psychiatric Association, and the American Psychological Association passed a joint statement requesting an increase in governmental protection and action in the form of v-chips, acceptable times for the airing of adult content, and a television rating system for shows (Joint Statement on the Impact of Entertainment Violence on Children). Meanwhile, grassroots organizations like the Parents Television Council (2005) and Turn off the Violence (2005) have also stepped in by applying pressure to the Federal Communications Commission and television stations to cancel certain episodes and shows that may be too violent.

During its original seven-year broadcast, two different episodes of *Buffy the Vampire Slayer* were canceled by the WB television network because they were thought to too closely resemble actual events of violence in schools. The first, and most poignant, episode, "Earshot," (Season 3, Episode 18) had Buffy temporarily hearing the thoughts of those around her. With this psychic ability, Buffy "overhears" someone planning to kill everyone at the school. After a frantic search, Buffy finds a lone student at the top of a school tower with a rifle—even though he was intending suicide, and it was the lunch-lady that was planning mass-murder (through poisoning). Temporarily canceled and later rescheduled, the episode had too much similarity between the gunman and perpetrators in actual school shootings. Later in Season 3 (in "Graduation Day, Part Two," another episode also had a delayed broadcast (Season 3, Episode 22). In this episode, students come together during a graduation assembly in a rare (for the show) mass attack to, collectively, protect their school from the Demon-Mayor.

In an interview on the *Regis and Kathie Lee Show,* the actress that played Buffy (Sarah Michelle Gellar) discussed the cancellation of "Earshot." Though she agreed that the timing of the episode, so closely following the Columbine shootings, was wrong, she also spoke out against its cancellation—that we need to face tragedies rather than turn away from them (Bonin). This position also reflects Joss Whedon's (the creator of *Buffy the Vampire* Slayer) well-known attitude towards the role of television and violence. It is possible that the lack of realism in a show like *Buffy the Vampire Slayer* not only reduces the effects of media violence, but augments positive messages of the consequences of violence as well. Interestingly, many fans protested the WB's decision by downplaying the violence by emphasizing the fantasy nature of the violence that perpetrated against evil demons (Zillmann, Dolf and James B. Weaver 145–165).

In fact, research shows that the fantastical nature of the violence reduces the negative effects of media violence on the viewer (Heller and Polsky 279–285). Other factors also reduce the negative effect of media violence on the viewer, such as the level of pain cues or harm to victim, and consequences to the perpetrator (Heller and Polsky 279–285; Kunkel et al., 284–291; Potter and Smith 301–323). What these different forms of violence have in common is that they reflect Bandura's findings of modeling violent behavior only when the original behavior has not been punished (1977). Though not yet studied, writers could show characters as negatively influenced by violence by changing the characters rhetoric or speech patterns after a violent act. For example, characters that exhibit lowered verbal complexity after committing a violent act may demonstrate that violence lowers intelligence, and that avoiding violence is preferable.

Violence, Rhetoric and Verbal Complexity

Previous research has shown that adolescents who exhibit physical signs of aggression and violence are characterized by lower verbal complexity (e.g., Cole 332–343; Dumas, Blechman, and Prins 347–358). Factors like word frequency, the number of different words, and the total number of words used define verbal complexity. Research on criminality has reported that children who perpetrate violent crime also tend to exhibit lower IQs, often defined by lower verbal skills (Petee and Walsh 1987). This link between violence and verbal complexity has lead researchers to include training in skilled information exchange for children who are at risk due to violence exposure (Dumas, Blechman, and Prinz 347–358).

A reflection of the association of lower verbal complexity and violence is a generic convention in television and film. Catchy comebacks and snappy quips are often paired with heroes and villains attacks. For example, in *The Terminator* the tag line of the movie is when the villain, played by Arnold Schwarzenegger, tells the heroine, "I'll be back." This comment occurs after two hours of gratuitous violence as the Terminator systematically kills and blows up property (all in an attempt to murder one woman). The creators of *Buffy the Vampire Slayer* often spoof this phenomenon of the snappy quip. For example, in Season 1, Episode 12 ("Prophesy Girl"), Buffy enters "The Master's" lair and states, "You know, you really oughta talk to your contractor. Looks like you've got some water damage." The Master replies, "Oh good, the feeble banter portion of the fight!" This pattern of exchange continues throughout the show. Before putting a stake through a vampire's heart in "Wild at

Heart" (Season 4, Episode 6) Buffy says, "You eat this late you're gonna get heartburn." As the vampire dies quickly and mutely Buffy exclaims in despair, "That's all I get, one lame-ass vamp with no appreciation for my painstakingly thought-out puns?" However, as popular and memorable as the some of these catch phrases and puns may be, the link between them and the violent rhetorical situation is not yet connected. If the verbal complexity of television characters accurately reflects the association between verbal complexity and violence seen with at risk populations, then characters should decrease in verbal complexity after engaging in violence (Cole 332–343; Dumas, Blechman, and Prins 347–358).

While much work has been done on slang in *Buffy* that research has not yet accounted for what the pairing of slang, puns, or clichéd language, and violence might activate for viewers.[2] In David Fritts' "Warrior Heroes: Buffy the Vampire Slayer and Beowulf," he states that "Buffy's use of language has been frequently associated with her power" and that "words fail Buffy when she is powerless" (20). Furthermore, "the fact that Buffy chooses punning and wordplay as her battlefield voice suggests control and detachment" (Fritts 24). In pairing concrete data with readings, such as Fritts,' we can provide another way in to analyzing violent rhetoric.

Consistent with the previously mentioned research, Buffy's verbal complexity may decrease when she engages in violence, linking lower verbal complexity and violence. This pattern could convey to the viewing audience that Buffy is somehow diminished after engaging in violence. Simply put, Buffy sounds unintelligent due to violence. On the other hand, increases in Buffy's verbal complexity when she engages in violence would be inconsistent with previous research, linking higher verbal complexity and violence. This pattern could convey to the viewing audience that when Buffy engages in violence, she is augmented, becoming smarter while violent. Following this, we examine the patterns in the relationship between verbal complexity and Buffy's rhetoric of violence.

Method

All seven seasons of *Buffy the Vampire Slayer* were coded for violent incidents. Violent incidents on the part of Buffy are divided into three intervals: pre-incident, incident, and post-incident. An *incident* is defined as beginning when Buffy first initiates physical contact with her supernatural adversary and ending when the adversary was killed by Buffy. The analysis was restricted to incidents in which Buffy kills an obviously supernatural adversary (e.g., vampire or demon) in order to remove the possibility of an ambiguous victory influencing the level of Buffy's verbal complexity. For this essay, verbal complexity is used to mean how linguistic complexity is reflected by the terms of the Flesch Reading Ease score, and we will return to this shortly. A lower number on this scale means a sentence that is easier to understand, and a higher score indicates a sentence with more knowledge and linguistic acuity. Thus, a lower score reflects lower complexity. The current effort was also restricted to incidents in which Buffy was the only "good" character to kill a supernatural adversary in order to remove any diffusion of responsibility for the killing from the audience's point of view. Finally, in order to hold the putative level of violence constant, the current effort was restricted to incidents where Buffy kills only one supernatural adversary. The timing counter of the DVD player defined the duration of the violent incident. *Pre-incident* is defined as beginning when Buffy first entered the scene and ending when Buffy first initiates physical contact with her supernatural adversary. *Post-incident* is defined as beginning when Buffy kills her supernatural adversary and ending when Buffy leaves the scene.

For this process, text files of episode transcripts were downloaded into Microsoft® Word 2000 and marked with the duration times of the pre-incident, incident and post-incident. These segments of the transcripts were then analyzed for verbal complexity. Specifically the dialogue of Buffy was subjected to Microsoft Word 2000's calculation of Flesch Reading Ease.[3,4] Simply put, the Flesch Reading Ease score accounts for linguistic complexity—a lower score indicates a sentence that is easier to understand, and a higher score is indicative of a sentence that requires more knowledge and linguistic acuity. Thus, a lower score reflects lower complexity in Buffy's speech. Also tabulated were the number of protagonists and antagonists present, the number of supernatural adversaries killed, and the number of total words spoken by Buffy for each interval.

A potential explanation for Buffy's verbal behavior during incidents could be the level of Buffy's physical exertions. In order to address this possibility periods where Buffy engaged in training exercises (but not actual physical violence) were also coded (using the above guidelines). Since Buffy does not engage in mortal combat during training, a *training incident* is defined as beginning when Buffy first initiates physical activity without the presence of a supernatural adversary, and ending when Buffy is no longer engaged in physical activity. Pre-training and post-training are defined as Buffy entering the scene before initiating physical activity, and until Buffy left the scene after stopping physical activity, similar to the violent incidents.

Results

For all seven seasons (144 episodes), Buffy engaged in 106 acts of violent incidents and 86 of them fit the selection criteria.[5] Cases are excluded from analyses if they were missing a pre-incident, or post-incident. This rendered 70 total cases of violent behavior for analysis. For these 70 cases, on average, Buffy spoke at a 2nd grade level. Specifically, these analyses show that while Buffy speaks the same number of words before and after a violent incident, her verbal complexity decreases after she engages in violence. Thus, we have a link between lower verbal complexity and violence.[6] Examples of how this works, rhetorically, are examined in the following sections.

Ancillary Analyses: Training

As stated, physical exertion could be a potential explanation for Buffy's verbal behavior rather than the violence that Buffy engaged in. To address this possibility, we analyze periods when Buffy engaged in physical exertion (training), but not violence. For all seven seasons, Buffy engaged in 21 acts of training incidents that fit the selection criteria. As in the violent cases, training cases are excluded from analyses if they were missing a pre-incident, incident, or post-incident. This rendered seven cases of Buffy engaging in physical training. For these seven cases, on average, Buffy spoke at a 2nd grade level. Specifically, these analyses show that while Buffy speaks the same number of words before and after training her verbal complexity does not change after she engages in training.[7] We can conclude, therefore, the reduction of verbal complexity after a violent incident does not seem to be due simply to Buffy's exertions when fighting.

In the episode "Band Candy" (Season 3, Episode 8), we see how physical exertion during training exercises does not change Buffy's verbal complexity. In *Band Candy*, the gang gears up for studying for the SATs and selling candy bars for a school fundraiser. The major twist in this episode comes when everyone realizes that eating the candy makes the adults act like teenagers. Later on we learn that in this episode Giles (Buffy's watcher) and Joyce (Buffy's Mom) have sex. Prior to any of that, Buffy has a training session with Giles early in the episode (6:21–7:18). Pre-incident in training (6:21– 6:46), Buffy says, "Ow!... Why do I put up with this?... Okay, you're just doing this to take funny pictures of me." The Flesch score here is 97.9. During the incident (6:47–6:59) Buffy states: "You ran out of new training ideas about a week ago, huh? Okay. Five, four, three, two, one." The Flesch score is 92.6, which is a decrease from the pre-incident score. However, the resolution of the training exercise shows the Flesch score rises again to 98.7. Post-Incident (7:00–7:18), Buffy states: "Thanks!... I can't. Mom's in hyper drive. She wants me home tonight. I told you.... I know, I know. She's out of control. Enjoy the candy!" As the above data supports, Buffy's verbal complexity in this scene (and in her training exercises) does not undergo a significant decrease after she engages in training. This can help us to conclude, therefore, that the reduction of verbal complexity after a violent incident does not seem to be due simply to the physical exertion of fighting.

Ancillary Analyses: Incident

In the present analyses the incident of violence has been regarded as a treatment condition, wherein Buffy's verbal complexity decreases following an "injection" of violence. However, while Buffy's verbal complexity decreases after the incident it is also possible that Buffy's verbal complexity increases during violent incidents, thereby glorifying the violent act. In order to address this possibility, Buffy's verbal complexity during the incident is analyzed.

The (a priori) analysis revealed that there was a significant difference between the number of words spoken during the incident and the number of words spoken pre- and post-incident.[8] However, there was a nonsignificant difference between Buffy's verbal complexity during the incident and Buffy's verbal complexity pre- and post-incident.[9] However, this study revealed that there was a significant difference in Buffy's verbal complexity from the pre-incident to the post-incident.[10] Explicitly, these analyses show that while Buffy speaks fewer words during the violent incidents her verbal complexity additionally decreases after she engages in violence, thus linking lower verbal complexity and violence. Therefore, Buffy's violence does seem to act as a catalyst that leads to a decrease in verbal complexity.

We see this trend exemplified in "Lover's Walk" (Season 3, Episode 8). In this episode, the gang finds out their SAT scores—and Buffy surprises everyone with her score of 1430. The plot centers around Buffy considering a future that might include college, while Spike the Vampire holds Willow captive in the hopes she will cast a spell to win Drusilla back. One of the incidents analyzed in this episode shows the pattern described by the data above. Towards the end of the episode (time marker 33:20), Buffy, Spike and Angel face off with Lenny and his gang. Pre-incident (33:20–33:45), Buffy's dialogue is, "You know, he was just leaving. Don't you start anything.... The guys are in trouble. We can't risk this.... Sorry. We're Leaving..." This dialogue shows a Flesch reading of 91.5. There is no speaking during the incident (33:46–34:35). Post-incident (34:36–35:27), Buffy's dialogue reads: "Go! I violently dislike you." The spoken words here have a Flesch reading of 52.0. This reading of pre and post-incident rhetoric and violence shows that Buffy's decrease in verbal complexity following her fighting.

This is seemingly paradoxical considering the stereotypes inherent in the good guy/bad guy binary. Commonplace attitudes would hold that so-called good guys/gals like our heroine, Buffy, would be agents of all around goodness as they save the day. In the case of Buffy, after she slays the bad guys (and therefore saves the day) her words do not live up to a sense of quality or "goodness" that we would expect from a heroine. This complicates the expectations for a good guy/gal particularly when we consider the data on how violence and verbal complexity actually work (and we will return to this point shortly).

Ancillary Analyses: Other

Finally, other variables deserve consideration as to whether or not they contributed to Buffy's decrease in verbal complexity after engaging in violence. To examine whether other variables may have contributed to Buffy's decrease in verbal complexity a verbal complexity differential score is created by subtracting the pre-incident verbal complexity score from the post-incident verbal complexity score. Thus, a lower score reflects the pattern in this study of decreased verbal complexity after a violent incident. "Other" factors include Buffy's maturation, the number of good guys or bad guys (protagonists and antagonists) in a given fight scene, and duration of fight scenes.

As we see, Buffy undergoes various changes throughout the seven seasons that lead to her increasing maturity. Therefore, Buffy's verbal complexity may have increased throughout the seasons. Specifically, Buffy's decrease in verbal complexity post-incident may be a product of Buffy's changing verbal complexity across seasons. However, there was no relationship between season and Buffy's decreasing verbal complexity.[11] Take, for example, the second episode of *Buffy* and the first episode of Season 4 when Buffy enters college. In "The Harvest" (Season 1, Episode 2), the Flesch score is 78.1 Post-Incident, and in "Freshman" (Season 4, Episode 1) the Flesch score is 79.8 Post-Incident. We can see that there is a negligible change between the Post-Incident scores even in the different rhetorical situations of high school versus college. In other words, a partial college education hardly changes Buffy's verbal complexity.

Secondly, accounting for good guys (protagonists) and antagonists (bad guys) in scenes allows for additional conclusions. Due to the nature of the selection criteria (i.e., Buffy killing the antagonist during the incident), Buffy is more likely to speak with antagonist pre-incident or during an incident, and more likely to speak with protagonists post-incident. Therefore, Buffy's verbal complexity may be a product of the number of protagonists or antagonists involved in the her fight scenes. Specifically, Buffy's decrease in verbal complexity post-incident may be due to the lack of antagonists in the post-incident. However, there was no relationship between the number of protagonists and Buffy's decreasing verbal complexity nor was there a relationship between the number of antagonists and Buffy's decreasing verbal complexity.[12]

An instance where we can examine the conflicted role of good guy/gal gone bad is in the case study of Faith—the Slayer gone rogue. We see her start as a protagonist and transition to an antagonist, and she is another site worth exploring for verbal complexity. In the episode "Bad Girls" (Season 3, Episode 14) Faith transitions from the role of a protagonist to an antagonist. Buffy's verbal complexity Post-Incident does not significantly change after Faith's transformation. This analysis could also be extended to other instances where characters take on roles of both good and bad—or even evil (as in the (plentiful) examples of Angel, Oz (when in werewolf-mode), Giles (when he is possessed), Spike, Willow-the-doppelganger, Buffy and the Buffy-bot, to name a few).

Finally, the more screen time Buffy had during each incident the more time Buffy had to talk. Therefore, Buffy's verbal complexity may be a product of the total duration of the pre-incident, incident, and post incident. Specifically, Buffy's decrease in verbal complexity post-incident may be due to an increase in the total duration of a scene. However, there was no relationship between the total duration of a scene and Buffy's decreasing verbal complexity.[13] We can return to the above-mentioned examples of the second episode of *Buffy* and the first episode of Season 4, when Buffy enters college. As explained previously, both episodes have similar Post-Incident Flesch scores. Both episodes also have largely different screen-time for their Incidents. In "The Harvest" (Season 1, Episode 2), the primary Incident has a duration of 02:27, while in "Freshman" (Season 4, Episode 1), the primary Incident has a duration of 00:06. We can see that there is not a notable change between the Post-Incident scores even in the different durations of the fights.

Implication of Results

The results of this research show that the verbal complexity of our heroine, Buffy, decreases after she engages in violence. This is consistent with previous research (Cole 332–343; Dumas, Blechman, and Prins 347–358) that has shown that adolescents who are violent are characterized by lower verbal complexity. We cannot attribute this pattern of results to maturation (or lack thereof) in season, the number of other good guys or bad guys present in the scene, or the duration of the scenes. The pattern of results does not reflect Buffy's activity level during the incident, making it more difficult for her to speak after exerting herself physically. However, Buffy was also engaged in strenuous activity during training exercises. The same pattern of decreased verbal complexity does not occur during training, suggesting more was occurring when Buffy was fighting than just physical exertion.

After establishing a pattern for Buffy's verbal complexity, the logical next step would be to discover how the audience reads this information.

Portraying Buffy as the intellectually diminished good gal after violent incidents might lead the viewing audience to disparage violence. Bandura showed that children who saw punishment of perpetrators of aggression were less likely to engage in aggressive behaviors. Other researchers have also suggested that presenting characters as negatively influenced by the violence they have committed may send the message that violent behavior is undesirable (Heller and Pollsky 279–285; Kunkel et al. 284–291; Potter and Smith 301–323).

Future research may show that manipulating the verbal complexity of violent protagonists or antagonists may also exert the beneficial effects of reducing the negative repercussions of viewing media violence. Finally, though Buffy was not one of the most educated characters on the show she was intelligent enough to make it through high school and enter college—with very high SAT scores. However, this analysis reports the disturbing conclusion that even at her best Buffy speaks little more than a 2nd grade level. While much work has been done on the role of education in the show there still has yet to be a study that examines how that education is demonstrated, for example in verbal mastery, in the actions of the characters.[14] This study only looked at the verbal complexity of the main protagonist, and Buffy is still only one character. As mentioned earlier, Fritts argues that "words fail Buffy when she is powerless" (20), and "Buffy chooses punning and wordplay as her battlefield voice suggests control and detachment" (24). We can perhaps refine these claims to state that violence causes words to fail Buffy, and her "battlefield" voice suggest the opposite of control (24).

Future research may also examine other good guys/gals that are portrayed as more intelligent (e.g., the librarian, watcher Giles, or Willow the computer whiz) during violent incidents. For example, Willow, attempts to recreate Buffy's fighting-rhetoric in "Anne" (Season 3, Episode 1), saying, "Hey there Big Boy?" to a demon. This is followed up with a comment about how "The Slayer always says" something that "disarms" demons, and that they had previously taken Buffy's punning "for granted."

Additionally, it is worth inquiring if episodes that refuse standard spoken language still maintain the idea that violence leads to decreased intelligence for good guys/gals. Examples worth further inquiry include "Hush" (Season 4, Episode 10), where there is an extended lack of dialogue; "Bad Beer" (Season 4, Episode 5), where Buffy talks like the archetype of a cave-person after drinking laced beer; or the infamous musical episode "Once More, with Feeling" (Season 6, Episode 7). For these episodes, it is worth exploring a methodology to account for intelligence outside of language.

In Gywn Symonds 2015 essay, "'Solving Problems with Sharp Objects': Female Empowerment, Sex and Violence in *Buffy the Vampire Slayer*," she states that in *Buffy*, "we are given the distance to consider what the violence is being used to do or reveal or tell" (8). Violence here reveals trends we can quantify, and further extend to application in real-world situations. The conclusions from this also allow us to ask questions about representations of the good guy/bad guy binary in relationship to violence. It does seem counter intuitive that good guys/gals should speak less intelligently after something like slaying or violence in the name of a greater good. However, the data in this analysis shows otherwise. Regardless, the cliché of punning being "the lowest form of humor," may in fact be true in the case of Buffy's rhetoric and violence. In other words, Buffy talks bluntly and carries a pointy stick. Puns abound in tandem with violence and in this process we see a slaying of verbal complexity.

Notes

1. On October 19–20, 2002, theorists from many different fields and countries attended an international conference entitled "Blood, Text and Fears" devoted to *Buffy the Vampire Slayer* wherein the program's treatment of rape, sado-masochism, feminism and violence were discussed. Since then, the field of so-called *Whedonology* and the exploration of the academic *Buffyverse* has grown exponentially. We can see this in texts such as U. Melissa Anyiwo and Karoline Szatek-tudor's 2013 text, *Buffy Conquers the Academy: Conference Papers from the 2009/2010 Popular Culture/American Culture Associations*, the long-running Slayage Academic conference, and the massive Buffyverse online bibliographies (such as http://slayageonline.com/EBS/buffy_studies/buffy_studies_by_discipline.htm).

2. Scholarship that focuses on slang and pun-pattern in *Buffy* includes: Michael Adams' *Beyond Slayer Slang: Pragmatics, Discourse, and Style in* Buffy the Vampire Slayer (2006), *Slayer Slang: A* Buffy the Vampire Slayer *Lexicon* (2003), and *Verbatim: From the Bawdy to the Sublime, the Best Writing on Language for Word Lovers, Grammar Mavens, and Armchair Linguists* and Bonnie Kneen's "'Add it up, it all spells "duh"': the language of *Buffy, the Vampire Slayer*" (17 August 2012).

3. This is also known as the Flesch-Kinkaid Reading Score. We choose to stick with the shortened phrase "Flesch Reading Ease" throughout.

4. The Flesch Reading Ease score is defined as: $206.385 - (1.015 \times$ (# of words / # of sentences)) $- (84.6 \times$ (# of syllables / # of words)).
From the Flesch Reading Ease score, a verbal complexity score was derived by subtracting the Flesch Reading Ease score from 101. If Buffy did not engage in any speech during the pre-incident, incident, or post-incident, then it is scored for verbal complexity as missing data.

5. To clarify, the criteria for an incident are that Buffy spoke, fought and was the only one to kill one, supernatural "bad guy," and then she spoke again—in that consistent order.

6. A one-way repeated measures Analysis of Variance (ANOVA) of words obtained pre-incident (M = 26.36), and post-incident (M = 20.71) revealed a nonsignificant effect of the number of words spoken pre-incident than post-incident, $F(1, 69) = 1.015$, $p = .1586$. A one-way repeated measures ANOVA of verbal complexity obtained pre-incident (M = 15.02), and post-incident (M = 8.84) revealed a significant effect of verbal complexity for incident, $F(1, 35) = 3.11$, $p = .0434$.

7. A one-way repeated measures ANOVA of words obtained pre-training (M = 30.29), and post-training (M = 18.29) revealed a nonsignificant effect of the number of words spoken pre-training and post-training, $F(1, 6) = .419$, $p = .2716$. A one-way repeated measures ANOVA of verbal complexity obtained pre-training (M = 20.15), and post-training (M = 19.03) revealed a nonsignificant effect of verbal complexity for training, $F(1, 3) = .003$, $p = .4801$.

8. A one-way repeated measures ANOVA of words obtained pre- (M = 26.36), during (M = 7.33), and post- (M = 20.71) incident revealed a significant effect of incident $F(2, 138) = 7.088$, $p = .0006$. Specifically, $F(1, 69) = 15.538$, $p = 9.59E-05$. A one-way repeated measures ANOVA of verbal complexity obtained pre- (M = 15.02), during (M = 12.92), and post- (M = 8.84) revealed a marginally significant effect of incident, $F(2, 28) = 1.836$, $p = .0891$.

9. $F(1, 14) = .000035$, $p = .4977$.
10. $F(1, 14) = 10.738$, $p = .0028$
11. $r(36) = .0673$, $p = .3447$.
12. The relationship between the number of protagonists and Buffy's decreasing verbal complexity does not exist: $r(36) = -.109$, $p = .2574$, nor is there a relationship between the number of antagonists and Buffy's decreasing verbal complexity, $r(36) = -.052$, $p = .3783$.
13. We see no relationship between the total duration of a scene and Buffy's decreasing verbal complexity, $r(36) = .106$, $p = .2633$.
14. Work that examines the role of education in the Buffyverse includes: Michael Betancourt's "Educating Buffy: The Role of Education in *Buffy the Vampire Slayer*" (1998), Adriana Estill's "Going to Hell: Placing the Library in *Buffy the Vampire Slayer*" (2007), Jodie A. Kreider and Meghan K. Winchell's *Buffy in the Classroom: Essays on Teaching With the Vampire Slayer* (2010), Kevin K. Durand's *Buffy Meets the Academy: Essays on the Episodes and Scripts as Texts* (2009), and Cynthia Jeney's. "If the Apocalypse Comes... Email Me: Or, All I Need to Know about Online Distance Ed, I Learned from *Buffy the Vampire Slayer*" (2002).

Works Cited

Adams, Michael. "Introduction: Beyond Slayer Slang: Pragmatics, Discourse, and Style in *Buffy the Vampire Slayer*." *Slayage: The Online International Journal of Buffy Studies* 5.4 (2006): 20. 12 September 2016.
_____. "Preface." *Slayer Slang: A Buffy the Vampire Slayer Lexicon*. New York: Oxford University Press, 2003.
_____. "Slayer Slang." *Verbatim: From the Bawdy to the Sublime, the Best Writing on Language for Word Lovers, Grammar Mavens, and Armchair Linguists*. Ed. Erin McKean. San Diego: Harcourt, 2001. 134–141.
American Academy of Pediatrics, the American Medical Association, the American Academy of Child and Adolescent Psychiatry, the American Psychiatric Association, & the American Psychological Association. *Joint Statement on the Impact of Entertainment Violence on Children*. Congressional Public Health Summit, July 2000. Print.
Anyiwo, U. Melissa, and Karoline Szate-Tudor, eds. *Buffy Conquers the Academy: Conference Papers from the 2009/2010 Popular Culture/American Culture Associations*. Cambridge: Cambridge Scholars Publishing, 2013.
Bandura, Albert. *Social Learning Theory*. Englewood Cliffs, NJ: Prentice-Hall, 1977.
_____. *Social Foundations of Though and Action: A Social Cognitive Theory*. Englewood Cliffs, NJ: Prentice-Hall, 1986.
Bandura, Albert, Dorthea Ross, and Shelia Ross. "Imitation of Film-Mediated Aggressive Models." *Journal of Abnormal and Social Psychology* 66 (1963): 3–11.
Betancourt, Michael. "Educating Buffy: The Role of Education in *Buffy the Vampire Slayer*." *Transylvanian Journal: Dracula and Vampire Studies* 3.2 (1998): 1–12.
Bierly, Mandy. "Buffy 101." *T.V. Guide* 19 October 2002: 6. Print.
Bonin, Liane. "School Daze." *Entertainment Weekly*. 25 May 1999. Web. 11 Sept. 2007.
Buffy the Vampire Slayer. Dir. Joss Whedon. Perf. Sarah Michelle Geller, Nicholas Brendon, Alyson Hannigan. Fox, 1997–2003. DVD.
Cole, Doris. "Narrative Development in Aggressive Boys." *Behavioral Disorders* 26 (2001): 332–343. 12 September 2016.
Curry, Vicki L. "The Violence at Thurston High School: Individual Characteristics of the Victims as Related to Their Post-Traumatic Responses." *Dissertation Abstracts International: Section B: The Sciences and Engineering* 28 March 2002: 5958. Web. 5 September 2015.
Drewniok, Malgorzata Gosia. "'You're Strong. I'm Stronger': Vampires, Masculinity and Language in *Buffy*." *Joss Whedon: The Complete Companion: The TV Series, the Movies, the Comic Books and More*. Ed. PopMatters.com, Mary Alice Money. London: Titan Books, 2012. 49–56. Print.
Dumas, Jean E., Elaine Ann Blechman, and Ronald J. Prinz. "Aggressive Children and Effective Communication." *Aggressive Behavior* 20 (1994): 347–358. Print.

Durand, Kevin K., ed. *Buffy Meets the Academy: Essays on the Episodes and Scripts as Texts.* Jefferson, NC: McFarland, 2009. Print.

Edwards, Ellen. "Children of a Fearful World: When Lockdowns and Anxiety Compete with the Three R's." *Washington Post* 24 October 2002. Web. 1 October 2014.

Estill, Adriana. "Going to Hell: Placing the Library in *Buffy the Vampire Slayer*." *The Library as Place: History, Community, and Culture*. Eds. John Buschman and Gloria J. Leckie. Westport: Libraries Unlimited, 2007. 235–250.

Fritts, David. "Warrior Heroes: *Buffy the Vampire Slayer* and *Beowulf*." *Slayage* 5.1 (2017, forthcoming).

Heller, Melvin S., and Steven Polsky. "Television Violence: Guidelines For Evaluation." *Archives of General Psychiatry* 24.3 (1971): 279–285. Print.

Huesmann, L. Rowell. "Psychological Processes Promoting the Relation Between Exposure to Media Violence and Aggressive Behavior by the Viewer." *Journal of Social Issues* 42.3 (1986): 125–139. Print.

Jarvis, Christine. "School Is Hell: Gendered Fears in Teenage Horror." *Educational Studies* 27.3 (2001): 257–67.

Jeney, Cynthia. "If the Apocalypse Comes ... Email Me: Or, All I Need to Know about Online Distance Ed, I Learned from *Buffy the Vampire Slayer*." *Kairos: A Journal for Teachers of Writing and Webbed Environments* 7.3 (2002). Web. 3 October 2015.

Jones, Gerard. *Killing Monsters: Why Children Need Fantasy, Super Heroes, and Make-Believe Violence*. New York: Basic Books, 2002.

Kneen, Bonnie. "Add It Up, It All Spells "Duh'": The Language of *Buffy, the Vampire Slayer*." *Oxford Dictionaries Online*, 17 August 2012. Web. 12 October 2015.

Kreider, Jodie A., and Meghan K. Winchell, eds. *Buffy in the Classroom: Essays on Teaching with the Vampire Slayer*. Jefferson, NC: McFarland, 2010.

Kunkel, Dale, et al. "Measuring Television Violence: The Importance of Context." *Journal of Broadcasting & Electronic Media* 39 (1995): 284–291. Print.

McDonald, Kathleen. "The More Things Change: Buffy and Angel Enact a Modern-Day Sentimental Novel." *Americanization of History: Conflation of Time and Culture in Film and Television*. Ed. Kathleen McDonald. Newcastle upon Tyne: Cambridge Scholars, 2011.

"Nielson Ratings for Buffy's Seventh Season: National Ratings for the Seventh Season." Insightbbwww. 4 November 2002. Web. 1 October 2014.

Paik, Haejung, and George Comstock. "The Effects of Television Violence on Antisocial Behavior: A Meta-analysis." *Communications Research* 21 (1994): 516–546. Print.

Parents Television Council. 2007. http://www.parentstv.org. Web. 18 October 2015.

Pearl, D., L. Bouthilet, and J. Lazar. "Television and Behavior. Ten Years of Scientific Progress and Implications for the Eighties." *National Institute of Mental Health*. Rockville, MD, 1982.

Petee, Thomas, and Anthony Walsh. "Violent Delinquency, Race, and the Wechsler Performance-Verbal Discrepancy." *The Journal of Social Psychology* (1987): 353–354. Print.

Potter, W. James, and Stacy Smith. "The Context of Graphic Portrayals of Television Violence." *Journal of Broadcasting & Electronic Media* 44 (2000): 301–323. Print.

"Psyche: Buffy Transcripts." *Transcripts*. 17 October 2002. Web. 18 October 2015.

Savage, Joanne. "Does Viewing Violent Media Really Cause Criminal Violence? A Methodological Review." *Aggression & Violent Behavior* 10 (2004): 99–128. Print.

Symonds, Gwyn. "'Solving Problems with Sharp Objects': Female Empowerment, Sex and Violence in *Buffy the Vampire Slayer*." *Slayage: The Online Journal of Buffy Studies* (26 May 2015). Web. 10 October 2015.

The Terminator. Dir. James Cameron. Perf. Arnold Schwarzenegger, Linda Hamilton, Michael Biehn. Hemdale Film, 1984. DVD.

Wood, W., F. Wong, and J.G. Chachere. "Effects of Media Violence on Viewers' Aggression in Unconstrained Social Interaction." *Psychological Bulletin* 109 (1991): 371–383. Print.

Zillmann, Dolf, and James B. Weaver. "Effects of Prolonged Exposure to Gratuitous Media Violence on Provoked and Unprovoked Hostile Behavior." *Journal of Applied Social Psychology* 29 (1999): 145–165. Print.

"I need an antiheroine"
Female Antiheroes in American Quality Television

SOTIRIS PETRIDIS

In recent years, there has been a massive production of shows that include female characters who can be categorized as antiheroes. These shows are part of the American Quality Television, a term that emerged between scholars in the mid–1990s and describes a new style of television. This essay will examine the representation of female antiheroes—the so-called antiheroines—in contemporary American quality television. More specifically, this essay will analyze the characteristics of antiheroines in American television and will focus on three distinctive examples of female characters from well-known shows: Carrie Mathison (Claire Danes) from *Homeland* (2011–present), Claire Underwood (Robin Wright) from *House of Cards* (2013–present), and Annalise Keating (Viola Davis) from *How to Get Away with Murder* (2014–present).

But before I examine how the antihero archetype became a trend in contemporary American television, first I have to examine what quality television is and how the shows have been affected. Traditional television has undergone substantial changes, particularly through new serial formats on premium television channels leading to the so-called "quality television series" (Klarer, 203). In other words, quality television is describing a new way of programming that is of higher quality due to its subject matter, style, and content. So in a way, quality television is the product of the post-network era in the United States (Imre 392). Subscription-based cable networks did not have face advertising pressure because of the quality of programming. Once viewers subscribed based on the quality of programming, suddenly "quality" drama shows began winning the vast majority of Writers' Guild, Directors' Guild, Screen Actors' Guild, and Emmy awards (Dunne 106–107). The starting point of the term "quality television" is set in 1996, when the book of Robert J. Thompson, *Television's Second Golden Age*, was published. Thompson argued that "quality TV is best defined by what it is not. It is not 'regular' TV" (13). According to Janet McCabe and Kim Akass, the characteristics that define the post–1996 television era are the changes in broadcast delivery, the new systems of production and distribution, the economic restructuring based on brand equity and market differentiation, and the rise to prominence of Home Box Office (HBO), whose tagline "It's Not TV. It's HBO" characterizes its quality marker (3). All these characteristics led quality drama to become a genre in itself, complete with its own set of formulaic characteristics (Thompson 16). As Lindsay Garrison points out, quality drama shows have caused celebration from viewers, critics, and academics, all of which recognize creators as television auteurs (160).

Of course, nowadays quality television is not a privilege of cable television networks. Broadcast networks challenged the quality producers and writers who were succeeding on cable to put quality dramas in development for themselves (Dunne 107). There are numerous examples of quality television that are a product of broadcast television, like *Lost* (2004–2010) from ABC or *Prison Break* (2005–2009) from Fox. The changes quality television brought in shows' narratives helped the creation of more complicated and three-dimensional characters. The distinction between good and evil became unclear, and so did the rule of the protagonist—either as hero or antihero. As Meghan Gallagher comments, "there has been a huge shift in focus over the past few decades, towards more academic and complex television content that mirrors cinematic style. Audiences no longer wish only to be entertained, but also challenged." So American television started to have antiheroes as lead characters of both economically and artistically successful shows. As Paula Berman comments, "The antihero as dramatic lead is an undeniable trend in television. The dark protagonist, a charismatic, problematic man who challenges social norms and looks damned cool doing it." But what exactly is an antihero?

The "antihero" is a common character archetype that has been around since the comedies and tragedies of Greek theater, and unlike the traditional hero, the antihero usually has a flawed moral character (Michael). Antiheroes are main characters in literature, film, or television that lack the admirable traits of a traditional hero, while they fill the function of protagonist in a story's format but cannot be considered heroes due to their questionable morality (Gallagher). According to Shane Dayton, there is a touch of darkness surrounding antiheroes and a sense of danger that does not go away, while a character in this role "performs acts that are generally deemed or thought to be heroic, but he/she will do so with methods, actions, manners, and intentions that are not so."

The antihero archetype has historically been dominated by males. Gladys L. Knight thinks that this is no surprise, as women were commonly portrayed in film as do-gooders—nice women who followed the law, while rarely appear in the complicated role of an antihero (35). Antiheroes were present in audiovisual works long before the emergence of television, with a lot of cinematic genres—like film noir, westerns, and sci-fi, depending on this type of character. Some of the iconic antiheroes of cinema are: Phillip Marlowe (Humphrey Bogart), Dirty Harry (Clint Eastwood), Travis Bickle (Robert De Niro), Michael Corleone (Al Pacino), and Han Solo (Harrison Ford).

The most famous examples of television antiheroes were usually male. Walter White, Dexter Morgan, Rick Grimes, Don Draper, Ray Donovan, Nucky Thompson, and Jax Teller are only some of the antiheroes that have drawn audience's interest over the last years. According to Gallagher, Tony Soprano from HBO's *The Sopranos* (1999–2007) is often cited as the first antihero of contemporary quality television, because he was easy to empathize with while being a morally reprehensible criminal. Soon, other networks produced their own shows based on a leading character that fit into the antihero archetype, making it the norm and not the exception of quality television.

Antiheroes and Female Representation

As it was mentioned before, antiheroes were mostly male characters, while female protagonists held the roles of a do-gooder or an antagonist. Sundi Rose-Holt argues that "male characters who are flawed and multidimensional (often with horrible acts of their own doing in their past) are called antiheroes, but women are called villains." Even if quality television brought three-dimensional and more structured characters, women were still represented as flat characters that either were totally good or totally bad. Gallagher comments:

> It seems that edgy, complicated characters who stray from the path of morality, are incredibly popular, as long as they embody stereotypes of patriarchal masculinity. Despite the overwhelming popularity of male antiheroes, female antiheroes are still few and far between. Representations of women in television are still catching up.

It is true that in recent years, we see that something is changing and there are more three-dimensional female characters that can be described as antiheroes, though still not as many when compared to their male counterparts. Trying to find a reason of why this happens, Brenda Weber comments that "If men are depicted as antiheroes, it's still something the audience can rally 'round. But if an artist, author or filmmaker creates women who are antiheroes, suddenly, they are misogynists and they've killed feminism" (qtd. in Sciullo). This is, of course, an extreme statement that has a grain of truth, but does not take into account the filmmakers that took the risk of creating an antiheroine and the positive reinforcement of feminism as a movement regarding women's representation.

These new female representations came as a result of earlier, successful feminist struggles for equal rights. The struggles included the opportunities for women to have both family and careers, and to live and love like men without giving up their femininity (Imre 391). Third-wave feminism, and subsequently the first battles that have won as a movement, is the main reason leading to the appearance of antiheroines in contemporary American television. Third-wave feminists have had the privilege of growing up with the freedoms that first and second wave feminists fought for, while they maintained a youthful, almost idealistic optimism about their position in society and what they are entitled to. They also possessed a limitlessly inclusive definition of feminism stating that from the soccer mom, to the career woman, to the stripper, to the scholar, the feminist can create herself (Barkman 90). In other words, women can be whatever they want to be and still fight for equality.

This kind of thinking influenced American television and its way of representing women. Now, television is filled with strong, experienced, and capable female characters that include attorneys, doctors, politicians, CIA agents, and judges. Some of them have families and some of them do not. But their emotional and behavioral structure is not flat and stereotypical. As Rose-Holt comments on this subject:

> the traditional binary that exists in stories about women is beginning to shatter. Women don't have to work in either/or tales anymore, performing as the old hag OR the beautiful siren; the madonna OR the whore; the distressed damsel OR the sexless workaholic.

All of these led to the emergence of the antiheroine archetype. Even if there are not as many examples of female antiheroes as their male counterparts, there are enough so we can identify their existence as a new television trend. In the beginning of the new millennium, quality television has embraced the antiheroine archetype, which according to Harriet Minter is a strong, determined and morally suspect woman forging ahead on her chosen path. Cristina Yang from *Grey's Anatomy* (2005–present), Olivia Pope from *Scandal* (2012–present), Patty Hewes from *Damages* (2007–2012), and Alicia Florrick from *The Good Wife* (2009–2016) are some of the most well-known examples of antiheroines. The rest of this essay will focus on three distinctive and well-structured examples of antiheroines in American quality television: Carrie Mathison, Claire Underwood, and Annalise Keating.

Carrie Mathison from Homeland

Homeland is a political thriller, created by Howard Gordon and Alex Gansa, and it airs on Showtime, a cable network. The show focuses on Carrie Mathison, a CIA agent working as a field operative in Iraq, and Nicholas Brody (Damian Lewis), a U.S. Marine Corps Scout Sniper who was held captive by Al Qaeda as a prisoner of war. When she was in Iraq, Carrie met an imprisoned CIA asset, and moments before his execution, he told her that an American prisoner of war was turned by Al Qaeda figure Abu Nazir (Navid Negahban). Mathison believed that Brody was the double agent the asset was talking about, and therefore a threat to the United States.

Even though there are two main characters, the show is structured around the life and actions of Carrie. In the first three seasons we follow both of the characters, but we primarily see Carrie's perspective. Thus, the audience mostly knows only what she knows. Also, Carrie can be proven as the protagonist of the show by the fact that Brody dies at the end of the third season. Carrie is a dynamic character and has a demanding job that requires a lot of energy. She works for the CIA and dedicates her everyday life to protecting her country. Of course, as an original antiheroine, Carrie has a dark side too. She was diagnosed with bipolar disorder, for which she secretly begins to take drugs supplied by her older sister, Maggie (Amy Hargreaves). Secondly and most importantly, even if she strongly believes that Brody is a traitor, she starts having feelings for him that develop to a romantic involvement.

As the show develops, we see more of Carrie's dark side. She gets pregnant with Brody's child right before he's killed. Then, in season four we see Carrie being on a mission in Islamabad, while she left her baby with her sister. In a video call, her sister tells her that "your daughter is one week older than last time we spoke," making clear that Carrie does not ask about her baby as often as she should. And when ordered to return from Pakistan, Carrie finds it hard to even pick up her own baby, something that implies that her maternal instincts are completely absent.

A scene that underlines her dark side and makes her the antiheroine is when she bathes her baby and she slips underwater. Viewers see that Carrie is confused and full of terror, but it seems that the audience is going to witness a possible infanticide. Even if she rescues her child, the idea that she was capable of letting her baby die creates a morally controversial character. Usually, motherhood is one of the main female stereotypical representations placing female characters as the do-gooder of the narrative. As Rebecca Feasey comments, "the contemporary media environment is saturated with romanticized, idealized and indeed conservative images of selfless and satisfied 'good' mothers" (3). If a woman is represented as a bad mother, or even worse as a possible child murderer, she becomes the villain of the narrative. We see that in antiheroines' narratives we can watch a female character be as morally controversial as Carrie, yet still be the protagonist of the show. Antiheroes have to be both morally unacceptable and yet, still reasonably likable, so they are often less in control than they believe and this helps to make the characters sympathetic (Gallagher).

Carrie has other good qualities that are centered in her professional life. This is the reason why this example of an antiheroine is so important to the television archetype that is emerging; she has a family and a newborn child, yet she chooses work over them. Until recent years, there used to be a double standard regarding gender representation: men were free to choose something over their family and still be likable, while female characters ought to be nurturing and have maternal instincts. Carrie is one of the first women depicted as a "bad mother" and still is able to be the person that the audience will identify with. She is smart, intelligent, and has all the moral flaws she needs to be called an antiheroine.

Claire Underwood from House of Cards

Netflix is one of the first extensively successful streaming services with on-demand viewing, while exclusively operating outside the time-slot format of traditional television. In February 2013, Netflix released its own original show, *House of Cards*, and broke all the conventions of traditional serial distribution by making available the entire first season at once—something they also did in all of the seasons that followed. This type of distribution benefits the narrative in creating more three-dimensional and well-structured characters because the episodic format has changed and the emotional arc of every character is more concrete than ever before.

House of Cards is a political drama from Netflix, created by Beau Willimon. The show is about Frank Underwood (Kevin Spacey), a Democrat from South Carolina's 5th congressional district and House Majority Whip who initiates an elaborate plan to get himself into a position of greater power, always with the help of his wife, Claire Underwood. Even if the show focuses more on Frank Underwood, Claire is also a basic character, while both of them are antiheroes. Both characters are trying to climb the ladder of success and power, and the audience follows them moving from Congress to White House, since Frank becomes the President and Claire, the First Lady.

The two characters are well-developed and have the characteristics of the antihero archetype. Claire is a fine example of how an antiheroine coexists with a male antihero. As Amanda Ciske comments, "Claire Underwood's character captures the complexity of an ambitious woman balancing the pursuit of power, marriage and her inner desires. What makes her a role model is her ability to double as feminine, yet sharp and decisive to get what she wants." She is next to the most powerful man in the United States and still manages to be more active than anyone could imagine. A comprehensive example of this statement is the whole storyline of season 4 in which Claire manages to be the running mate of Frank as his Vice President. Roth Cornet describes her as the person who will stop at nothing to conquer everything. In the first season she is depicted as a lobbyist who runs an environmental nonprofit organization. Also, she has an extramarital affair with an artist friend, something that, as in the case of Carrie, makes her break the conventional rules of a "traditional" family. Beau Willimon, the creator of the show, says that "what's extraordinary about Frank and Claire is there is deep love and mutual respect, but the way they achieve this is by operating on a completely different set of rules than the rest of us typically do" (qtd. in Oldenburg).

In season two, she uses her interview where she admitted that she was raped in college, and that her rapist is now a high-ranking general, to lobby for support for legislation supporting victims of military sexual assault, and then adapts the focus on that issue into political support to her and Frank. Over this storyline, the audience witnesses her dark side that makes her an antiheroine, because it sees her intentionally ruin the lives of her ex-lover, the First Lady and a fellow rape victim, while threatening the well-being of a pregnant former employee of hers (Dockterman).

In season three, when she becomes the First Lady, Claire feels the need to be something more noteworthy than this, and asks Frank to nominate her to a United Nations position. Even though the Senate rejects her, she demands the job and takes it by Frank in a recess appointment. The First Lady is stereotypically linked with a more passive representation, which involves her full support to the President. Also, the representation of wives in television is similar, with Bernie Kanner to note that the women usually are less prominent and "they revered their husbands as more important than themselves" (48). In the case of Claire, we see that not only does she want something more than that, she demands it and attains it while causing political cost to her husband. Claire Underwood is a fine and well-structured example of a contemporary television antiheroine. She is ambitious, active and determined to get what she wants. Even if she is next to a man who has the same thirst for power, she still manages to get what she deserves, no matter the cost.

Annalise Keating from How to Get Away with Murder

How to Get Away with Murder is a legal drama, created by Peter Nowalk, which airs on ABC. The show is about Annalise Keating, a law professor at a prestigious Philadelphia university who, with five of her students, becomes entwined in a murder plot. The first season of the show explores two important murders through flashbacks—Lila Stangard (Megan West), lover of Annalise's husband, Sam Keating (Tom Verica). Even though ABC is a broadcast network, this show is clearly part of quality television; it is edgy, dark, and has an antiheroine as a leading character. The quality of the show can be proven by its many awards, including the Emmy Award for Outstanding Lead Actress in a Drama Series, which made Viola Davis the first African American woman to win in this category.

Annalise is the latest and finest example of an antiheroine in quality television. She is a perfectionist, manipulative, workaholic woman who has power over the rest of the characters. She has both good and bad sides, making her a well-structured and a three-dimensional antiheroine, with Davis to note that Annalise is the most complex character she's ever played (Rosen). Nowalk, the creator of the show, said: "I wanted to create someone who was enigmatic and controversial and dark. I think people are much darker than we give them credit for on TV" (qtd. in Dos Santos).

There are many examples in both her personal and professional life, which prove that Annalise is an antiheroine. She helps her husbands' killers to avoid being caught, and she is willing to sacrifice the freedom of her lover, Nate Lahey (Billy Brown) by framing him to the police. She did so with the help of Frank (Charlie Weber) by hiding Sam's wedding ring in the woods and planting Nate's fingerprints on the evidence. So, Annalise is willing to send an innocent man, who used to be her lover, to jail to save her students.

Apart from her personal life, Annalise is ruthless in her job, too. In the first episode of season two, Annalise and her team try to steal a case about two adopted siblings accused of murdering their parents from another lawyer. She does so by assigning her students to humiliate their current representation by modifying footage to discredit the main witness. And with this unethical and illegal action, she manages to take the case away from her colleague.

Annalise is not an ordinary antiheroine because she pushes the boundaries of the archetype. She not only breaks the convention of the male antihero, she also breaks the norm of the white antihero. Kelsey Wallace comments:

> As a black woman, Annalise Keating packs just as potent a combination of sex appeal and unlikability as Tony Soprano or Don Draper with the added bonus of not being a white dude. She dominates her professional life and her personal life, and is respected and feared by everyone who crosses her path.

Annalise breaks every rule of the white male antihero archetype, because she is not only a woman, but she is also an African American. And this is not the only pioneering element of this character because she is part of a show from a broadcast network. As it was aforementioned, quality television, which made antiheroes a trend, started from cable television and after the big success of these shows, broadcast networks started to produce their own in order to catch up. Annalise is a fine example of how a show can be revolutionary and innovative by using an existing television archetype, while pushing its boundaries.

Conclusion

American quality television regenerated the medium and reinvented the norms of the audiovisual narrative. This change started from cable television and then spread to broadcast networks. One of its innovative norms was the establishment of the antihero archetype as a leading character of a show. Quality dramas were more dark and obscure than the television shows of previous years. As Jonathan Michael comments, "brokenness is a part of humanity, and we can more easily relate to the choices that a character makes on a TV show if they are broken too. After all, a believable and relatable character is one of the single-most important elements of an enjoyable story."

For many years, the antihero archetype was standardized and the representation of a white male character was almost stereotypically reproduced. In the last years, a lot of shows experimented with this norm and changed the representation of the classical antihero. So, the first antiheroines made their appearance. Feminism had a crucial role in this change, because the freedom of being who you are and still be a feminist was used as a norm in the archetype.

This essay analyzed three distinctive examples of contemporary television antiheroines; one from cable television, one from online streaming service, and one from broadcast television. All three women share certain attributes, while each of them breaks several stereotypical norms of women's representation in media. Antiheroines are becoming to be a central part of quality television and one thing is certain: they are going to concern us for a long time.

Works Cited

Barkman, Ashley. "And the Followship Award Goes to … Third-Wave Feminism?" In *30 Rock and Philosophy*, edited by J. Jeremy Wisnewski. Hoboken: Wiley, 2010. Print.

Berman, Paula. "The Female Antihero on Television. Sisters Are Doing It for Themselves." *Broad Street Review*, 31 May 2014. Web. 25 Sept. 2015.

Bernie Kanner. "From Father Knows Best to The Simpsons: On TV, Parenting Has Lost Its Halo." In *Taking Parenting Public: The Case for a New Social Movement*, edited by Sylvia Ann Hewlett, Nancy Rankin, and Cornel West. Lanham: Rowman & Littlefield, 2002. Print.

Cornet, Roth. "Netflix's Original Series House of Cards—From David Fincher and Kevin Spacey—May Be the New Face of Television." *IGN*, 31 Jan. 2013. Web. 2 Oct. 2015.
Dayton, Shane. "Top 10 Greatest Movie Anti-Heroes." *Listverse*, 14 Feb. 2008. Web. 22 Sept. 2015.
Dockterman, Eliana. "9 Most Shocking Moments from House of Cards Season 2." *Time*, 17 Feb. 2014. Web. 20 Aug. 2015.
Dos Santos, Kristin. "Why Viola Davis Is the Least Likeable Heroine Ever in a Shonda Rhimes Show." *E Online*, 25 Sept. 2014. Web. 3 Oct. 2015.
Dunne, Peter. "Inside American Television Drama: Quality Is Not What is Produced, but What It Produces." In *Quality TV: Contemporary American Television and Beyond*, edited by Janet McCabe and Kim Akass. London: I.B. Tauris, 2007. Print.
Feasey, Rebecca. *From Happy Homemaker to Desperate Housewives: Motherhood and Popular Television*. London: Anthem Press, 2012. Print.
Gallagher, Meghan. "Scandal's Olivia Pope and the Rise of the Female Antihero." The Artifice, 5 Jan. 2015. Web. 30 Sept. 2015.
Garrison, Lindsay H. "Defining Television Excellence 'On Its Own Terms': The Peabody Awards and Negotiating Discourses of Quality." *Cinema Journal* 50:02 (2011): 160–165. Print.
Imre, Anikó. "Gender and Quality Television." *Feminist Media Studies* 9:04 (2009): 391–407. Print.
Klarer, Mario. "Putting Television 'Aside': Novel Narration in House of Cards." *New Review of Film and Television Studies* 12:02 (2014): 203–220. Print.
Knight, Gladys L. *Female Action Heroes: A Guide to Women in Comics, Video Games, Film, and Television*. Santa Barbara, CA: Greenwood, 2010. Print.
McCabe, Janet, and Akass, Kim. "Introduction: Debating Quality." In *Quality TV: Contemporary American Television and Beyond*, edited by Janet McCabe and Kim Akass. London: I.B. Tauris, 2007. Print.
Michael, Jonathan. "The Rise of the Anti-Hero." Relevant, April 26, 2013. Web. 27 Sept. 2015.
Minter, Harriet, Amanda Ciske et al. "What We Can Learn from TV's Leading Anti-Heroines." *The Guardian*, 18 Feb. 2015. Web. 15 Sept. 2015.
Oldenburg, Ann. "House of Cards Promises More Plotting and Scheming." *USA Today*, 13 Feb. 2014. Web. 15 Sept. 2015.
Rose-Holt, Sundi. "Antiheroines Are the New Antiheroes: The Killer Women of 'Penny Dreadful,' 'Orphan Black' and More." Indiewire, 3 June 2015. Web. 1 Oct. 2015.
Rosen, Lisa. "'Murder's' Annalise and Other TV Women Leap Darkly into Antihero Scene." *Los Angeles Times*, 18 Dec. 2014. Web. 23 Sept. 2015.
Sciullo, Maria. "Strong, Fascinating Women in TV Dramas Often Saddled with Quirks to Make Them More Interesting." *Pittsburgh Post-Gazette*, 28 Jan. 2015. Web. 29 Aug. 2015.
Thompson, Robert J. *Television's Second Golden Age: From Hill Street Blues to ER*. New York: Continuum, 1996. Print.
Wallace, Kelsey. "'How to Get Away with Murder' Gives Us a Female Antihero We Can Root For." *Bitchmedia*, 26 Sept. 2014. Web. 29 Aug. 2015.

A Dangerous Mind
An Examination of the Effectiveness of Walter White as Educator

Richard L. Mehrenberg

Breaking Bad creator Vince Gilligan has often described his pitch for the show in a single sentence. In an interview with National Public Radio, he called it "the story of the transformation of a man from Mr. Chips to Scarface" ("Breaking Bad"). This brief synopsis highlights the path of a desperate man who went from being a beloved school teacher to a vicious criminal. Anyone with a passing familiarity with the show can immediately recognize and appreciate the analogy's endgame. There are numerous parallels between Walt White and drug kingpin Tony "Scarface" Montana. Both characters are steeped in a culture of drugs, power, money, and violence. Both characters are self-made men with a thirst for recognition, appreciation and respect. Finally, both men are undone by their own hubris and sense of invulnerability.

Saying that Walt White is good at making meth is like saying that Mozart was good at making music. His extreme knowledge, talent, and skill allowed him to rise to the elite stratosphere of his field. By using any yardstick to measure success (e.g., $80 million cash, a product of 99.1 percent purity, the complete eradication of all competitors and antagonists), Walt White proves that he earns the title of "the twenty-first century Scarface." It remains more ambiguous as to whether Walt is also worthy of the first half of Gilligan's analogy. Was he a skillful and effective teacher? Does he deserve comparison to Mr. Chips?

A bit of background information might be helpful in describing the character of Mr. Chips, who may have faded into relative obscurity for many familiar with only current popular culture. "Goodbye, Mr. Chips" was a 1934 novella written by James Hilton. It was later adapted into numerous films, stage plays, television series, and other media. The protagonist of the story is Mr. "Chips" Chipping, a beloved teacher at an English Boys' boarding school. Mr. Chips was respected as a very knowledgeable, beloved, and effective educator over the course of his long career. In generations past, his name was used as complimentary shorthand to describe a teacher of the highest quality.

It is noteworthy that Gilligan would reference such an admirable and noble educator like Mr. Chips, since *Breaking Bad* often portrayed the teaching profession as one of minimal value and respect in the eyes of society. Walt teaches adolescents who are, for the most part, uninformed and uninterested in chemistry. There seems to be a significant, palpable sense of boredom in his classroom, with both the teacher and students "going through the motions" until the bell rings.

Likewise, the tangible rewards for teaching are seen as minuscule or non-existent in the *Breaking Bad* universe. Walt is required to take on a part-time job at the local car wash to make ends meet. He initially decides to cook meth because his school health insurance doesn't cover his medical bills. He uses his life savings (less than $7,000) to purchase a run-down RV for his initial cooks. It is unclear whether these facts represent Gilligan's opinion of teachers, how he perceives society views teachers, or a combination of the two.

Regardless of the source of perceived negativity, there is ample evidence to make an informed judgment regarding the quality of Walter White as an educator. By the examination and interpretation of key dialogue, actions, and events from the series, it can be deduced that the great Heisenberg falls far short of the educational prowess of Mr. Chips.

To judge the effectiveness of Walt as a teacher, we need some sort of criteria or benchmark. In real life, it is an extremely difficult task to quantify the effectiveness of teachers. Most current teacher evaluation systems are criticized for being either too narrow or too arbitrary in their criteria (Danielson 35; Hull 22). Tools such as formal observations are prone to subjectivity. Standardized student test scores only measure a very narrow window of achievement and are prone to bias and questionable validity. Long term outcomes such as the percentage of students who get full time jobs or graduate from college are mired in confounding variables like socioeconomic status, geographic location, and parental level of education.

The ongoing need for effective instruction has served as the catalyst for at least one group, to define, identify and reward exceptional teaching on a national level. In 1987, a nonpartisan, non-profit organization named the National Board for Professional Teaching Standards (NBPTS) was established to provide an objective and rigorous path to recognize and reward seasoned educators within 25 subject areas and developmental levels. Through a year-long process of submitted portfolio entries, video-taped lessons, and a formal series of essay exams, licensed teachers with a minimum of three years of experience submit materials to demonstrate the required knowledge and skills associated with a master teacher.

Over the past twenty-five years, nearly 100,000 teachers have successfully completed the process. A substantial amount of academic research has been dedicated to demonstrating that National Board teachers provide a higher level of effectiveness and student achievement when compared to their non-certified peers (Salvador, Samantha, and Baxter 8). The level of effectiveness is most prominent within low-performing schools (Cavaluzzo 20; Goldhaber and Anthony 145).

Regardless of subject area, all National Board Certified Teachers are required to demonstrate mastery of five core propositions. These basic tenets serve as the cornerstones for what an accomplished educator is supposed to know and be able to do. The applicant's portfolio entries and essay exams are directly connected to the five propositions. They are:

1. Teachers are committed to students and their learning.
2. Teachers know the subjects that they teach and how to teach those subjects to students.
3. Teachers are responsible for managing and monitoring student learning.
4. Teachers think systematically about their practice and learn from experience.
5. Teachers are members of learning communities.

What follows is an examination of each of the five core propositions as they apply to the pedagogical effectiveness of Walter White.

Teachers Are Committed to Students and Their Learning

National Board Certified Teachers strongly believe that all students can learn. They easily recognize students' strengths and interests and are able to align instruction accordingly. They are able to create personalized incentives for students as appropriate.

In his high school chemistry classroom, it is not clear that Walt always abides by these beliefs. A conversation in the pilot episode illustrates that his former student and current meth partner, Jesse Pinkman, did not find success in his class.

> WALT: Did you learn nothing from my chemistry class?
> JESSE: No. You flunked me, remember? ["Pilot"].

In another episode, Jesse pulls out one of his old chemistry tests tagged with a large red F and Mr. White's somewhat harsh comments, "Ridiculous! Apply yourself!" However, during the second season episode "Grilled," Jesse's mother is interviewed by Walt's brother-in-law/DEA agent Hank Schrader. Mrs. Pinkman speaks positively of Walt's influence on Jesse as his former educator. "Mr. White must have seen some potential in Jesse. He really tried to motivate him. He was one of the few teachers who cared" ("Grilled"). These examples suggest that Walt White has very high expectations of his students. He holds students accountable for learning. Grades in his classes are earned, not gifts.

Although having high expectations for students is a desirable trait of effective teachers, it does not necessarily guarantee a belief that all students can learn or that student strengths and interests play a large part in instruction. In fact, it may suggest the opposite. It may suggest a certain pedagogical stubbornness, a "my way or the highway" mentality. This is sometimes referred to as the "Pontius Pilate" effect. This type of teacher presents a lesson and puts all expectation for comprehension on the student. He "washes his hands" of students' failing grades and poor tests scores (Dufour and Eaker 58). Educators who regularly employ this approach do not truly engage the student or put forth the effort required for interactivity. Instead, they simply serve as a resource of information, no different than a dictionary or internet search engine.

To get a more thorough understanding of Walt's approach to the first core proposition, we should look at the specific way that he interacts with Jesse beyond the high school classroom. There are many examples throughout their partnership for which Walt belittles Jesse's intelligence and competence. At one time or another, he has called him an "imbecile," "moronic," a "junkie loser," and used the terms "Einstein" and "genius" sarcastically. Such activities do not positively reflect on Walt's commitment to his student.

As the series progresses, Jesse's intellectual growth evolves considerably. He shows insight, creativity, and problem-solving skills that equal or even exceed Walt's. For example, in the season five episode "Live Free or Die," Walt and Mike are unable to figure out a way to access incriminating computer evidence locked away in the police storage unit. It was Jesse who suggested using magnets as a way to erase the hard drive. A few episodes later, "Dead Freight" finds Walt and Mike stumped again, and Jesse devises a brilliant plan to steal methylamine from a train without resorting to violence, or even detection.

It could be expected that Walt would shower praise, or at the least acknowledgment towards his protégé for these flashes of brilliance. Sadly, Walt remains aloof in his response. In both instances, Walt is not seen praising, or even positively acknowledging, his pupil for his efforts. It seems as if he is unable or unwilling to celebrate the intellectual accomplishments of a student. Walt attempts to motivate Jesse mainly through insults and humiliation. He accentuates the negative, and ignores the positive. This educational philosophy is considered unproductive, if not detrimental in the classroom. The evidence suggests that Walt does not strongly adhere to the first core proposition.

Teachers Know the Subjects That They Teach and How to Teach Those Subjects to Students

National Board Certified teachers have a solid understanding of their content. Many have advanced degrees and extensive real-world experiences that serve as the backbone to their pedagogy. Accomplished teachers also understand that a mastery of the content is not enough. They must seek out effective and meaningful ways to convey the material to students. One of Walt's greatest strengths as educator is his knowledge of chemistry. He has an extensive educational background that includes graduate school at the California Institute of Technology (CIT), a research-intensive private university renowned for its innovative scholars and cutting-edge scientific applications.

Walt also has substantial real-world past experiences that enhance his credibility as a master chemist. He even earned a certificate of recognition from his former company, Gray Matter, for helping them to win a Nobel Prize for proton radiography. This prestigious accomplishment belies his humble work as a high school chemistry teacher.

Although it is indisputable that Walt knows the subject he teaches, it is questionable whether or not he knows how to teach this subject to students. From the point of view of a drug manufacturer, it could be argued that only a master teacher would be able to successfully instruct Jesse how to replicate his "Blue Sky" crystal meth recipe. While working for Gus Fring, Walt taught Jesse how to synthesize a complex amalgamation of ingredients to create an extremely pure and highly desirable product. He also had to instruct Jesse how to utilize sophisticated tools and complex machinery to ensure near perfection in the end result. This is not a bad accomplishment for someone who flunked high school chemistry!

Upon further reflection, it becomes less clear if Walt is an extremely effective teacher, or if Jesse is an extremely motivated student. When Walt later mentors Todd, it quickly becomes apparent that his new protégé is not up to the task. Walt recommends that he "take notes" and "pay attention," but these overly-simplistic suggestions appear laughable to the regular viewer who has previously witnessed the enormity and complexity associated with a Blue Sky meth cook. It comes as no surprise that Todd fails at being able to replicate the recipe on his own.

Quality teaching goes beyond simple recommendations to "take notes" and "pay attention." It involves elaboration, clarification, and discussion. *Breaking Bad* provides us with several examples of how Walt's lack of elaboration or clarification leads to serious problems. For example, in the episode "Cat's in the Bag…," Walt requires Jesse to get a very specific type of plastic container for the disposal of a drug dealer's corpse. However, he fails to elaborate to him why this specific type of plastic container is so important. When Jesse is unable to find the exact plastic container, he improvises by dumping the corpse and the powerful acids into his upstairs bathtub instead. The corrosive mess eventually eats through the tub and the bathroom floor below it. A disgusting soupy-red mess plummets through the hallway ceiling. Walt witnesses this and sarcastically turns to Jesse and says, "I'm sorry, what were you asking me? Oh, yes, that stupid plastic container I asked you to buy. You see, hydrofluoric acid won't eat through plastic; it will however dissolve metal, rock, glass, ceramic. So there's that."

It is both surprising and disappointing that a veteran high school chemistry teacher would neglect to explain the potentially lethal consequences associated with ignoring his orders ahead of time. Instead, Walt's hubris causes to him strike out with sarcasm and arrogance ex post facto.

Similarly, in the episode "4 Days Out," while in the middle of the desert, on a marathon cooking spree, Walt reprimands Jesse for leaving the keys to the RV on the work station (aka-counter). Walt is afraid that Jesse will lose the keys and that they will end up stranded. Jesse is instructed to put the keys "in a safe location." In this instance, at least Walt has the foresight to predict a possible problem (lost keys). Unfortunately, he does not follow through and suggest an appropriate alternative location for the keys. Jesse put the keys in the RV ignition which accidentally drains the battery. This leads the pair to become stranded. Walt responds by humiliating Jesse through a series of insults and verbal attacks.

The accomplished teacher needs to be familiar enough with both his student and his content that he can recognize common challenges and potential misunderstandings. He knows how to explain content in a clear and patient manner. He can explain not just "what" to do, but also "why." When problems arise, he seeks out alternative ways to teach the material, rather than blame the student. He never resorts to sarcasm or verbal abuse. Walt has significant difficulties meeting many of these criteria.

Teachers Are Responsible for Managing and Monitoring Student Learning

Master educators have transparent lesson objectives that are directly aligned with assessment. Quality teaching limits surprises. Students know what is expected of them, what they need to accomplish, and the specific criteria that will be used to judge their level of competency. There are no hidden agendas.

Walt White is not very straightforward when it comes to his goals. He regularly uses manipulation and deception to get what he wants. This is especially problematic when it comes to interactions with his student, Jesse. For example, in season three episode "Green Light," Jesse starts cooking meth on his own. When Walt learns of this, it is clear to the viewer that Walt is secretly threatened by Jesse. In the following conversation, Walt attempts to use reverse psychology to cast doubt in Jesse's recipe:

> WALT: This is very shoddy work, Pinkman. I'm actually embarrassed for you.
> JESSE: What? No way. I gave out samples, and everyone said it was the bomb.
> WALT: Oh, they said it was the bomb? And who are they, I wonder? A bunch of meth heads?
> JESSE: Yeah, and they should know, right?

By belittling his efforts, Walt uses manipulative statements of disappointment and shame to mask his intimidation and vulnerability. His "objectives" and his "assessment" do not match up.

Another example of extreme manipulation occurs in the episode "End Times," when Jesse is about to shoot Walt over poisoning Brock. Walt pleads his case and convinces Jesse (and the viewer) that Gus Fring is behind the attempted murder of a child.

> Walt: I have been waiting.... I've been waiting all day, waiting for Gus to send one of his men to kill me, and it's you. Who do you know who's OK with using children, Jesse, who do you know? Who's allowed children to be murdered, hm? Gus! He has been ten steps ahead of me at every turn and now the one thing that he needed to finally get rid of me is your consent, and boy he's got that down, he's got it. And not only does he have that, but he manipulated you into pulling the trigger for him.
> JESSE: But only you and I knew about the ricin!
> WALT: No! You don't even believe that. Gus' cameras everywhere, please. Listen to yourself. No, he's known everything all along. Where were you today? In the lab? And you don't think it's possible that Tyrus lifted the cigarette out of your locker? C'mon! Don't you see? You are the last piece of the puzzle. You are everything that he's wanted. You're his cook now. You're the cook and you have proven you can run a lab without me, and now that cook has reason to kill me. Think about it! It's brilliant! So go ahead, if you think that I am capable of

doing this, then go ... *[Walt grabs Jesse's wrist and puts the gun on his own forehead]* ... put a bullet in my head and kill me right now.

Jesse is then convinced that Walt is innocent of the crime, and agrees to team up with him to destroy Gus Fring and his meth empire. Later, Jesse finds out that Brock was never actually exposed to ricin but instead a highly toxic houseplant called "Lilly of the Valley." The *Breaking Bad* viewers are also convinced of Walt's innocence until we are privy to a surreptitious camera shot of the potted plant located in Walt's backyard. At this point, we learn that we have been manipulated and lied to as well.

Accomplished teachers must be transparent in their interactions with students. By masking their true intentions, not only is instruction weakened, but rapport and positive relationships developed with students will diminish, and eventually evaporate. What teachers say and what they do must be in tandem.

Teachers Think Systematically About Their Practice and Learn from Experience

There is always room for improvement; the master instructor knows that he can never fully master his craft. The teacher is also a student, and must continually develop his skills. He recognizes his imperfections and learns from his mistakes.

Walt is repeatedly sucked into a world of violence, danger, and death. After overcoming a treacherous situation by the skin of his teeth, he resolves, time and time again, that things will be different. They never are. For example, after the violent deaths of drug dealers, Krazy 8 and Emillio, Walt needs to convince Jesse to go back into the meth trade with him. He tells him that he has learned from their mistakes, and guarantees their safety with an empty promise. "No matter what happens, no more bloodshed. No violence" ("Crazy Handful of Nothin'").

In the episode "Breakage," after the death of drug kingpin, Tuco Salamanca (which involved considerable bloodshed and violence), Jesse and Walt bicker over the possibility of going into business for themselves. In this instance, Walt is able to acknowledge mistakes from the past, yet he does not learn from them.

> JESSE: I remember, oh, I remember. That you cook, I sell. That was the division of labor when we started all this. And that's exactly how we should have kept it! 'Cause I sure as hell didn't find myself locked in a trunk or on my knees with a GUN to my head before your greedy old ass came along, alright?
> WALT: Alright, I will admit to a bit of a learning curve.

> JESSE: Oh-ho!
> WALT: And perhaps I was overly ambitious. In any case, it's not gonna happen that way anymore.

Walt fails to learn from his mistakes with other characters, too. In an episode, ironically titled "No Mas," Walt finally admits to Skyler that he cooks meth. He resolves to turn over a new leaf, and leave his criminal past behind. He arranges a meeting with Gus Fring to (temporarily) decline a lucrative offer to continue cooking for him.

> WALT [to Gus]: I'm here because I owe you the courtesy and respect to tell you this personally. I'm done. It has nothing to do with you personally. I find you extraordinarily professional and I appreciate the way you do business. I'm just…. I'm making a change in my life is what it is, and I'm at something of a crossroads and it's brought me to a realization: I'm not a criminal. No offense to any people who are, but … this is not me.

A few episodes later, after quickly giving in to the temptation, Walt confesses to Gus how his actions have negatively impacted his relationship with his family, especially his wife. "I have made a series of very bad decisions and I cannot make another one." All of these instances illustrate Walt's inability to learn from his mistakes. The old quote about "the definition of insanity is doing the same thing over and over again and expecting different results" is especially appropriate in this situation. Walt's hubris prevents him from recognizing that there are conditions in life that are completely beyond his control. His genius intellect and his over-confidence can only get him so far.

The most famous example of Walt's overconfidence occurs during an argument in the season four episode "Cornered," between Skyler and him. Skyler has become justifiably anxious and paranoid regarding the safety of her family. She confronts Walt and asks him if they are in danger.

> WALT: "No, you clearly don't know who you're talking to, so let me clue you in. I am *not* in danger, Skyler. *I am the danger!* A guy opens his door and gets shot and you think that of me? No. I am the one who knocks!"

"I am the one who knocks" has entered the lexicon alongside other Hollywood badass quotes like "Go ahead, make my day," "Say hello to my little friend," and "Hasta la vista, baby." The sentiment running through each is that the hero asserts his dominance and control over a powerful adversary.

Some may argue that Walt is accurate in his assessment that he *is* the danger. After all, he is able to eliminate all of his enemies (Gus, Hector Salamanca, Mike, Uncle Jack, Lydia). With the exception of being roughed up by Mike, very little physical harm is ever inflicted on him by his adversaries. When he is eventually shot and killed, it stems from his own doing, on his own terms. On the surface, he appears to be in control.

However, upon closer examination, one is reminded that the safety and well-being of Walt's loved ones are destroyed as a result of his actions. Hank and his partner, Gomie are murdered. He loses the love and respect of his wife and son, and will never get to know his baby daughter. His family is emotionally, financially and socially devastated by his choices. Furthermore, Skyler, the kids, and their home are all placed in extreme jeopardy due to Jesse's arson attempt and Todd's breaking and entering.

Walt has a brief, yet seemingly sincere, moment of honesty and repentance in the series finale. Before he embarks on his suicide mission against Uncle Jack and his crew, Walt goes to say goodbye to Skyler and Holly. He wants to explain to his wife why he made the choices that he did. Skyler fully expects to hear the same bogus rationale that Walt has provided since the very beginning, that everything he did, he did for the benefit of his family. Instead, Walt admits the real reason. "I did it for me. I liked it. I was good at it. And, I was really ... I was alive" ("Felina").

Walt's admission demonstrates that it is possible for him to learn from his mistakes. By admitting to his wife, why he did what he did, he gives her a certain sense of closure. Perhaps an earlier admission may have changed the fate of both him and his family.

Teachers Are Members of Learning Communities

The accomplished teacher values collaboration. He recognizes and appreciates the talents, skills, and resources of other educational professionals. He understands that even the best teachers have limitations on their knowledge and abilities, yet through working alongside them in formal (e.g., professional organizations) and informal (e.g., the teachers' lounge) environments, he is able to provide superior opportunities for his students.

Does Walt White "play nice with others"? He has a thorough history of forming, and then destroying partnerships in his evolution as a drug lord. His primary and initial collaborative relationship is, of course, with Jesse. Walt knows that Jesse has the knowledge, experience and connections to get him started in the drug trade. It must be noted that right off the bat, there is no sense of equity between partners. Walt threatens to blackmail Jesse by turning him over to the drug enforcement agency if his former student does not go along with his plans. Jesse agrees reluctantly; Walt gets what he wants.

From then on, Walt seeks out and exploits a string of professional relationships with criminals to advance his level of power, wealth, and notoriety. Walt comes into each of these relationships (Tuco, Gus, Mike, Lydia, Uncle Jack) at a distinct disadvantage. The other individual has significantly more power, wealth, experience, or resources than Walt. However, Walt is able to exploit each situation to get exactly what he wants (i.e., money, supplies, customers, mercenaries). The end result of each partnership is death. Specifically, Walt murders four out the five individuals after they outgrew their usefulness, or stood in the way of him achieving his goals. Therefore, it can be safely surmised that Walt engages in partnerships until he gets what he wants, at which point, he terminates the relationship, often literally.

In a separate, but equally compelling example of Walt's refusal to collaborate is his exodus from Gray Matter. Although the specific details are never fully explained, Walt walks away from a business promising unlimited prestige and wealth rather than collaborate with individuals with which he could no longer get along. By abandoning Gray Matter, Walt walks away from his share of a company eventually worth $2.6 billion. This represents a sum that would require over twenty five storage units to contain.

It can be argued whether or not *Breaking Bad* suggests that crime does not pay. After all, Walt ultimately accomplishes many of his self-centered goals, mostly on his own terms. Regardless of the verdict, one thing is for sure. The Gray Matter sub-plot proves that playing by the rules *does* pay … handsomely.

Conclusions

Using the five core propositions of the NBPTS as a lens, it becomes clear that Walt White would not be considered a master teacher. His key strengths include a deep understanding of his content matter, and substantial formal and real-world educational experiences. Walt demonstrates, time and again, that he is a brilliant chemist.

However, there is a large chasm between knowing the content and having the skills necessary to convey information to students in an accessible and supportive manner. Walt is mediocre at being able to explain chemistry in a way that "clicks" with those who do not share his level of knowledge or enthusiasm. Walt often shows impatience when his student does not understand or makes a mistake. He does not have a very well-defined pedagogical "bag of tricks" for struggling learners. Beyond telling them to "take notes" and "pay attention," he has little to more to offer.

Walt also resorts to sarcasm, verbal abuse, and humiliation. Such behaviors are extremely counter-productive to learning, and severely undermine the student-teacher relationship desired by effective educators. Furthermore, persistent belittlement of students would lead to the eventual probation or dismissal of the offending teacher.

The analysis of the effectiveness of Walter White as a teacher may be fodder for a more holistic critique of the entire series. Teachers have been historically considered among the most noble, admired, and socially respected professions in our country. Being an excellent teacher not only requires exceptional knowledge, skills, and experience, but also the appropriate set of dispositions. Was Walter's transformation into Heisenberg truly an evolution, or did he always have devious intentions slightly below the surface waiting to emerge? Perhaps, our antihero had very little "bad" to "break" in the first place.

Teaching is hard; expert teaching is even harder. It requires a great deal of patience, understanding, skill, and practice. Not everyone can do it, nor does everyone want to. Walt White may be the twenty-first century Scarface, but he was *never* Mr. Chips.

Works Cited

"About Us." *NBPTS*. National Board for Professional Teaching Standards, 19 Sept. 2014. Web. 23 Sept. 2015.
"'Breaking Bad': Vince Gilligan on Meth And Morals." *NPR*. National Public Radio, 19 Sept. 2011. Web. 23 Sept. 2015.
Cavalluzzo, Linda. "Will National Board Certification of Physical Educators Improve the Quality of Teaching?" *Journal of Physical Education, Recreation & Dance* 74.2 (2004): 18–20. *NBPTS*. Web. 23 Sept. 2015.
"Crazy Handful of Nothin'." *Breaking Bad*. AMC. 2 March 2009. Television.
Danielson, Charlotte. "Evaluations That Help Teachers Learn." *The Effective Educator* 68.4 (2010): 35–39. Print.
DuFour, Richard, and Robert Eaker. *Professional Learning Communities at Work TM: Best Practices for Enhancing Students Achievement*. Solution Tree Press, 1998.
Goldhaber, Dan, and Emily Anthony. "Can Teacher Quality Be Effectively Assessed? National Board Certification as a Signal of Effective Teaching." *The Review of Economics and Statistics* 89.1 (2007): 134–150. Print.
"Grilled." *Breaking Bad*. AMC. 15 March 2009. Television.
Hull, Jim. "Time to Do It Right." *American School Board Journal* (2013): 22–23. Print.
"Pilot." *Breaking Bad*. AMC. 20 January 2008. Television.
Salvador, Samantha, and Andy Baxter. "National Board Certification-Impact on Teacher Effectiveness." Charlotte-Mecklenburg Schools, 1 Oct. 2010. Web. 23 Sept. 2015.

Building (an) Empire
Queer Sacrifices in Lee Daniels' Empire

Robert LaRue

> I sing the body electric,
> The armies of those I love engirth me and I engirth them,
> They will not let me off till I go with them, respond to them,
> And discorrupt them, and charge them full with the charge
>   of the soul.
>   —Walt Whitman, "I Sing the Body Electric" (1885)

Wrapped within Whitman's "Body Electric" is, without a doubt, a celebration of equality. It is a celebration of the physical, of the body, over the abstractions of metaphysics; it is a pledge to make physical that which risks falling into abstraction; it is a promise to enjoy and revel in the messy physicality of one's world. More than this, though, within the poem exists a remarkable desire for unity: unity of self with others, unity of a national body, and unity of the inside and the outside. Yet, this unity is built on an unyielding urge to belong to another and to belong to oneself—a sense of completeness. M. Jimmie Killingsworth has asserted the poem, along with the general corpus of *Leaves of Grass*, as "provid[ing] a kind of manifesto on the political power of sex" (Killingsworth 125). I agree. I, however, would also like to add that in signifying the body as electric, Whitman forces us into a relationship with the conservation of energy itself: energy (or the body) is neither created nor destroyed; it is transferred—that is, it changes forms or undergoes transformation. In other words, if Whitman teaches us anything it is to look for and appreciate unity that is not predicated upon pure bodies but is constituted only through the collapsing of forms, of figures, of beings.

    I begin with Whitman because he represents perhaps *the* quintessential American "bad boy" image. As Christopher Beach asserts, "Whitman attacks the class system on which he believes the European" origins of American morality have been founded—especially within the context of literature (Beach 73-74). Moreover, I begin with Whitman because, as Killingsworth contends, Whitman "helped to invent gayness" by helping to develop "a rhetoric with the resonant power of an established discursive formation," thus creating ways of speaking about previously unnamed and unnamable experiences, feelings, and acts (122). Most importantly, however, I begin with Whitman because I find the irony of his now venerated position to be instructive for handling contemporary queer politics. Condemned in his own time for his "affront upon the recognized morality of respectable people" (qtd. in Killingworth 126), the once "bad boy" Whitman is now praised as a saving angel of a contemporary American gay identity for helping to create a gayness that refused to cleanse itself of those parts society found objectionable. Contemporary children of the Adamic Whitman, however, seem to have misinterpreted his democratic message as a call for assimilation. To a large extent, a politics of innocence, built from a rejection of that which society finds odious, has become the tactic of choice for contemporary lesbian and gay (LGBT) movements. Desires to make LGBT bodies more acceptable are translated as imperatives to depict queer bodies as clean and desexualized, and as close to being as "normal" as possible. Many more radical LGBT members, preferring the term "queer," have taken to criticizing this turn towards innocence. Lamenting the cleansing of queer sexuality, which, for them, holds *the* potential for uprooting systems of, what Adrienne Rich has termed "compulsory heterosexuality," these critics have offered important interventions in how queers (no matter where they fall on the political scale) are understood. However, this criticism directs the majority of its attention on the effects of the turn to innocence on the liberatory potential of queerness itself.

This turn to innocence becomes all the more problematic when considered in light of mainstream re-presentations of gays and lesbians. To construct a presentable image for mainstream society, "the gay community," as Essex Hemphill points out, "operates from a one-eyed, one-gender, one-color perception of 'community' that is most likely to recognize blond before black, but seldom the two together" ("Introduction" xix). This is to say rather than working to fulfill Whitman's call for a democratic body that longs for an equality of bodies, "the gay community" sacrifices black (queer) bodies, presenting them as excess and deviant, as an attempt to purify queerness. While the act and effects of "white-washing" gayness has been discussed, the causes for such behavior have not been given sustained attention. Therefore, adding to the objections raised by queer scholars, I would like to suggest an oft uncommented upon cause: I propose the "white-washing" of queerness as the result of a turn towards a politics of innocence as a tactic for gaining acceptance. When employed as a tactic in an ongoing battle for belonging, innocence becomes anything but innocent. Because it demands sacrifice, the tactic of innocence institutes violence akin to the violence of anti-gay sentiment. What I aim to demonstrate is that innocence can never be a strategy for acceptance or inclusion. I do not build this statement on the premise of a lost or denied pleasure, but on the recognition that certain bodies (not simply actions) are always excluded from the very formulation of innocence itself.

While, most immediately, this is an essay concerned with a politics of innocence which has been used by contemporary LGBT movements, resting at its core is the question of belonging—a question of origins and return. More directly, I am particularly interested in the role innocence plays in creating a space for belonging. It seems that, when dealing with race and/or queerness, discussions, more often than not, lead back to questions of belonging, whether by way of questions of origins (the origin of a black culture; the origins of categories of race; the origins of queer identities; etc.) or of a desire to return (return to a tradition of pride; return to a history of self-love; etc.). I state this, perhaps obvious, point not as a means of criticizing these questions, but as a means of arriving at an alternative vantage point.

To tease out the complications caused by a politics of innocence, I focus on the character Jamal Lyon (the queer, middle son of Lucious and Cookie Lyon) from the first season of the television drama, *Empire*. While some, especially within the black community, lament the show's repetition of the tropes of black people as drug dealers, loud, and hyper-sexual (see for instance, Curtis Bunn's "5 Reasons Why"), *Empire* has been lauded for placing black bodies in the mainstream media in some very important ways. Centered on the lives of the Lyon family—Lucious (the patriarch), played by Terrence Howard; Cookie (the matriarch), played by Taraji P. Henson; Andre, played by Trai Byers; Jamal, played by Jussie Smollett; and Hakeem, played by Bryshere Y. Gray—the show weaves black popular culture and the entertainment industry with black politics and black social issues, and helps foreground often suppressed conversations, such as homosexuality and mental illness within the black community (as aforementioned, Jamal is gay; additionally, Andre suffers from bipolar disorder).

On the surface, the show follows the Lyon family as they battle for control of Lucious' empire, aptly titled "Empire Entertainment," as well as the opportunity to become a power-player within the U.S. marketplace. The driving tensions within the show are twofold, yet inseparable. On the one hand, there is the drama produced by the struggles between family members—Luscious vs. Cookie; Andre vs. Jamal and Hakeem; Hakeem vs. Jamal, etc. On the other hand, there is the drama produced by Lucious' struggle to turn his company into an ubiquitously present, publicly-traded brand that does not just produce, but dominates the popular music scene. Each family struggle is in some way intensified by the struggle for the company and the power it promises, while the struggle for the company is a struggle for the family and its legacy.

Yet at the show's core rests Lucious' desire to mold his sons into what he sees as strong (black) men via lessons he learned from his own rise to power. Although he has great hopes for his sons, his desire to shape them constantly positions him as an antagonist against each son at various times. Lucious regrets that Andre, the degreed businessman, is not "street" enough, that he lacks the grit necessary to make difficult decisions. And he fears that Hakeem, the youngest son who came of age after Lucious' own acquisition of fame and wealth, possesses the street smarts Andre lacks, but lacks the necessary work ethic. And then there is Jamal, the son who possesses both Andre's business smarts and Hakeem's street smarts, but who is openly gay. Of the three sons, Lucious recognizes himself in Jamal more than any of the others. Jamal is smart and desires to rise and be recognized on his own merits, through his own talent and hard work. However, Lucious' desire to make (black) men of his sons causes him to reject Jamal because he chooses to live openly as a gay man. In Lucious' eyes, masculinity and Jamal's desire to live openly as a gay man are incompatible. Nonetheless, as the show progresses, Lucious comes to see Jamal differently—but not because he learns to accept Jamal's queerness.

Tracing Jamal's path to acceptance by his father, I suggest that the acceptance Jamal finally receives only becomes possible as he shifts away from a politics of innocence. I argue that Jamal's transition demonstrates the unsustainability of innocence as a tactic for acquiring acceptance, while also making visible how race and gender are sacrificed to maintain the concept of innocence.

Physicality: Queer vs. Gay

The shift of contemporary LGBT politics from concerns of sexual and social revolution to institutional (e.g. marriage and military) acceptance have received much critique by the more radical proponents of queer studies. Michael Warner, for instance, argues that, in an attempt to combat the shame and stigma so often attached to gay identities, the contemporary gay and lesbian movements have shifted towards a politics of innocence, which seeks to eradicate those elements of homosexuality that might be offensive or repugnant to heterosexual American sensibilities. Warner warns that "the distinction between stigma and shame makes it seem as though an easy way to resolve the ambivalence of belonging to a stigmatized group is to embrace the identity but disavow the act.... Dignity on these terms is bound to remain inauthentic" (33). Following a train of thought akin to Warner's, Lee Edelman urges queers, and queer studies, to embrace their abject position—even at the cost of futurity. Edelman finds that "the ethical burden" of the queer is "to figure an unregenerate, and unregenerating, sexuality whose singular insistence on jouissance, rejecting every constraint imposed by sentimental futurism, exposes aesthetic culture—the culture of forms and their reproduction, the culture of Imaginary lures—as always already a 'culture of death' intent on abjecting the force of a death drive that shatters the tomb we call life" (Edelman 47–48). In short, turning away from the current trend in LGBT politics seeking to assert the (domestic and social) normalcy of homosexuality, scholars like Warner and Edelman insist on the power of deviant and non-reproductive notions of queerness.

However, despite queer studies' resistance to attempts that seek to paint homosexuals as "just like" heterosexuals, many queer people of color (POC) continue to lament the field's lack of attention to the ways in which "national recognition and inclusion ... is contingent upon the segregation and disqualification of racial and sexual others from the national imaginary" (Puar 2). Ironically, as it attempts to (re)insert sex into the politics of (homo)sexuality, asserting that sex offers the most productive means of destabilizing systems of normalization, since it "is disruptive and aberrant in its rhythms, in its somatic states, and in its psychic and cultural meaning" (Warner 55), queer theory resorts to its own form of a politics of innocence. Often, as it repositions the true victims of society as those individuals who refuse to help "build a movement of homosexuals without sex" (48), it elides the ways in which race affect not only the "politics of shame," which Warner posits as the tactic de jour of contemporary LGBT movements and queer theory's attempt to liberate sex. In other words, the racial dynamics grounding the difference between notions such as "the closet" and the "down low" are not seriously interrogated by either contemporary LGBT *or* queer movements. Rather, both sacrifice discussions of race in order to privilege their brands of innocence.

In his essay "Outside in Black Studies: Reading from a Queer Place in the Diaspora," Rinaldo Walcott suggests black queer studies as a way of "rejuvenat[ing] the liberatory moments of the black studies project" (91). Arguing for that by thinking the "queer unthought" within black studies, an exploration of black queerness can help mitigate the field's dependence on notions of homogeneity. Recognizing that such a dependence on homogeneity arises from the field's simultaneous effort to counter educational and pedagogical wounds, which have left a gaping hole in the place of black contributions to society, and to manage the pressures of institutionalization, Walcott, nonetheless, urges a rethinking of the limits of homogeneity. According to Walcott, "we must confront singularities without the willed effort to make them cohere into a oneness; we must struggle to make a community of singularities of which the unworking of the present ruling regime, a regime that trades on the mythos of homogeneity, must be central" (93). Walcott's rejection of homogeneity resonates with the argument made in this essay. Within Walcott's reading of homogeneity, notes of a politics of innocence can be heard. When it is points out that "it is only too obvious to say that by and large the black studies project has in its thought produced [the] black community as assumed and essentially heterosexual" (Walcott 92), there are echoes of Frantz Fanon's desire to inoculate colonized black males from the "neurotic homosexuality" of white Europeans (Fanon 180, fn144), and Molefi Asante's assertion that "homosexuality is a deviation from Afrocentric thought because it makes the person evaluate his own physical needs above the teachings of national consciousness ... [and] threatens to distort the relationship between friends" (Asante 57). Such formulations of a homosexual-free black community come from men attempting to resist institutionalized readings of black (male) bodies and behaviors as perverted and deviant. In this sense, a flat reading of them as homophobic is too simple—after all, the entire foundation of their work is premised on the love of black men (and women). A more productive reading of such assertions might be to read them as attempting to apply a political strategy of innocence.

Attempting to redraw the boundaries of queerness, Cathy Cohen indicts both the "the liberal, rights-based agenda that dominates LGBTQ politics today" and contemporary queer movements that fail to "undress the contradictory relationship of marginalized folks, often communities of color, toward normalization" (Cohen 128, 131). Likewise, Roderick Ferguson proposes a queer of color critique to "shed light on the ruptural components of culture, components that expose the restrictions of universality, the exploitations of capital, and the deception of national culture" (Ferguson 24). For scholars of color such as Puar, Cohen, Ferguson, and Michael Hames-García race marks an important, and scarcely analyzed site of difference within a queerness that remains remarkably white. As I will endeavor to make clear in the discussion that follows, Jamal's transition from a good guy to a bad boy, and the subsequent acceptance he gains from that shift, demonstrates not only the limits of a politics based on innocence, but also the complicated relationship queers of color have with the largely white, mainstream LGBT and queer movements.

Unity: Jamal Lyon: A Portrait, of Sorts

In many ways, Jamal Lyon, as he is initially introduced to viewers, stands as the poster child for contemporary LGBT movement—and yet, ironically, the furthest from Whitman's democratic view. Jamal is tall, handsome, and in possession of a level of sophistication and refinement lacking in his two brothers. It is he, after all, who refers to Lucious' proposal to have his sons vie for control of his company, "Empire Entertainment," as an iteration of Shakespeare's *King Lear* (Daniels and Strong, "Pilot").[1] And unlike his brothers, Jamal is, initially, genuinely disinterested in battling for control of the company. As he himself claims, "[Lucious will] never pick me anyway ... [there is] way too much homophobia in the black community" ("Pilot"). "I," he continues, "don't want it anyway, so whatever" ("Pilot"). What he desires is for Lucious to "get to know me ... [to] spend some time with me" ("Pilot"). Jamal goes so far as to confess to Lucious that as a child he would think, "if I could make music like him, there's no way in hell that this man [Lucious] wouldn't love me" ("Die but Once"). It is at the behest of Cookie that Jamal takes up the challenge to prove himself as a viable candidate. Yet, the most important component of Jamal's initial presentation, and the quality that most likens him to contemporary LGBT movements and their politics of innocence, is the absence of any real "dangerous" sexuality. After recognizing the possible complications to his candidacy—his aforementioned queer sexuality and the homophobia within the black community, which is to say Lucious' homophobia—Michael, Jamal's boyfriend, ever so gently turns Jamal's face towards him and kisses him ("Pilot"). Then the telephone rings. The second instance of Jamal's sexuality, which finds him and Michael in bed, under their white sheets, is no more passionate than the first ("The Outspoken King"). This point is emphasized as Cookie strolls into the room without being the least bit shocked by what she encounters and then casually lays herself across the bed ("The Outspoken King"). While perhaps being initially shocking for their direct presentations of queer sexuality on mainstream television, these instances lack any real or threatening passion—that is, the type of passion and threat queer scholars and activists long to see embraced. In fact, the scenery for each instance is soaked in white light, suggesting the purity of Jamal and Michael's affections and actions. This is all the more significant given that these are two of the few overt expressions of Jamal's adult sexuality in the season.[2] As viewers are introduced to, and made to fall in love with him, Jamal is presented as the one character of innocence against a background of lies, deceit, and immorality that structure the Lyon family, and (the) *Empire* itself.

While I do not go so far as to suggest that "gay political groups owe their very being to the fact that sex draws people together and that in doing so it suggests alternative possibilities of life" (Warner 47–48), I do suggest that the cleanliness of innocence offers no room for the messiness of queer attachments and/or queer passions. In place of a passionate queer sex-life, Jamal substitutes a passion for music and artistic creativity. This sacrifice is all the more poignant when considered against later (that is, postlapsarian) depictions of his sexuality. While earlier depictions of Jamal's sexuality are devoid of the passion one would find visualized in the (hyper)sexual(ized) escapades of *How to Get Away with Murder*'s queer character, Connor Walsh (a character who might be said to be the poster child for queer studies, as he defiantly and unabashedly embraces both the dark and dirty sides of his queerness), later depictions of Jamal's sexuality find him freely—and quite passionately—initiating sex in the office and other, non-bedroom, locales.

Though frustrated by Lucious' rejection of his homosexuality, Jamal clings to the hope that he can win his father's approval if he just becomes the kind of musician Lucious is recognized as being. Initially, this hope inspires Jamal to resist Lucious' demands to remain in the closet. Ignoring Lucious' protests and outright demands, Jamal openly lives with Michael, and decides to come out as a gay musician. The disclosure of his planned "coming out" ignites an argument between the father and son, resulting in Lucious issuing Jamal an ultimatum: stay in the closet or lose the financial support he has been receiving ("The Outspoken King"). While Lucious' ultimatum initially causes Jamal to waiver, causing Jamal to unexpectedly back out of a publicity shoot Cookie staged for him, it is after the following exchange between Lucious and Jamal that Jamal enters his darkest moments:

> LUCIOUS: I tried to tell you since you were a baby that it's not about black eyes or bloody noses in this world, it's life or death. And if you don't toughen up, these streets will eat your ass alive.
> JAMAL: Since I was a baby, you beat me. You told me that was to toughen me up. That was a lie. You beat me because you hate me. And you always will because I'm always gonna be who I am.
> LUCIOUS: I don't hate *you*. I don't know you. I didn't bring any women into this world, and to see my son become somebody's bitch, I don't *understand* you.
> JAMAL: You don't have to understand me. You don't have to understand me or have anything to do with me. I'm a man. A man. So you can keep ... your money, and whatever else that Lucious Lyon thinks that he owns. My obedience is no longer for sale ["The Devil Quotes Scripture"].

It is after this exchange that Jamal declares, for the first time, "I'm going after his empire. Imma take it" ("The Devil Quotes Scripture"). This declaration is important because it is the first time Jamal directly expresses an interest in (the) Empire.[3] More importantly, though, it also serves as an important point in Jamal's shift towards darkness. Framed in a close-up, with Jamal staring directly into the camera, Jamal is no longer bathed in the white light of his earlier scenes. Instead, on a wet and empty street, he is tainted by the yellow of the streetlights and the darkness of the night. Jamal finally comes out during his performance of a cover of one of Lucious' songs, altering the lyrics, "it's the kind of song that makes a man love a woman" to "it's the kind of song that makes a man love a man" ("The Lyon's Roar").

No longer receiving financial support from Lucious, Jamal and Michael move into a small apartment in an impoverished neighborhood. Moving into this space, and using studios outside of the ones his father owned, Jamal begins his process of becoming the man his father has become—and closer to the man his father desires him to be. Just like his father, his path requires him to travel into darker territories—not only of the city but of himself, as well. As he settles into the rundown neighborhood he begins to divest himself of the sheltered life he formerly enjoyed. He ends his relationship with Michael, on the grounds that Michael cannot appreciate the dedication Jamal needs to fulfill his career dreams, and his relationship with Hakeem, the brother to whom he is closest fractures. The collapse of these relationships functions as an important point of departure for Jamal. Both Michael and Hakeem, in their own ways, serve as Jamal's ties with his sense of innocence, as they help create a space of comfort and acceptance not offered by his father and his father's demands.

However, it is the moment Lucious and Jamal have a heart-to-heart via a music showdown that Jamal begins gaining that which he sought most: Lucious' understanding, if not acceptance. During the episode "Die but Once," Jamal, playing the piano, and Lucious, playing the guitar, approach a mutual understanding of one another as they battle to express their frustrations and fears to create a duet for Lucious' upcoming music event. During the song, ironically titled "Nothing to Lose," not only does Jamal learn why Lucious defines manhood in the way that he does—because he grew up in the streets, at a time when being perceived as weak could cost a man his life and everything he loved—but Lucious discovers the "grit" that he has long desired to see in his son. The song is dark and dangerous, and Jamal grows darker with each chord he plays and with each line he offers as a rejoinder to Lucious' very own. Then something surprising happens: when Jamal sings, "how the hell did my back get against the wall?" the song flips from Lucious as lead to him as backup for Jamal. To Jamal's verse, Lucious rejoinders, "get your back up off the wall" ("Die but Once"). From this duel forward, the two leave with a newfound sense of common ground, and with Lucious having a newfound sense of respect for his son—though not a full acceptance of him as a gay man.

More than evidencing the seductiveness of Lucious' immorality, as might be assumed, Jamal's transformation might be more effectively read as demonstrating the problematics of assuming innocence as a tactic for acceptance. Rather than allowing him to follow Whitman in his call to "discorrupt" other bodies, Jamal's politics of innocence find him in need of corruption. Much like Warner contends of contemporary LGBT movements, Jamal's new darker persona *and* his initial presentation of innocence are both formed as responses to the image of his father's manhood and life. Angered by his father's rejection, and before he capitulates to the dark manhood his father desires of him, Jamal professes to Lucious that "the songs that I'm writing will only further prove that I'm you, but on steroids" ("The Devil Quotes Scripture"). Ironically, as reactions, neither Jamal's darker side, *nor* his previous persona of innocence afford him much creative space. According to Robert Reid-Pharr, "the black individual who does not have some understanding, however vague, that they might reject commonsense notions of identity, that they might perhaps re-create themselves into something unknown or perhaps unimaginable, is an individual who lacks full status as a subject" (*Once You Go Black* 113). Because it fails to offer new and imaginative ways to re-create himself, Jamal's innocence, then, turns out to be not so innocent, after all. Rather, his innocence becomes yet another form of policing, thus constraining the expression of his Self.[4]

As Jamal begins his journey towards finding the acceptance he longs for, he highlights a crucial point of tension between queers of color and LGBT and queer movements: where queer and gay and lesbian theories rarely analyze the role of the family unit in queer identities, for queers of color, the family unit is often an indispensable component of one's identity. In their important essay "True Confessions: A Discourse on Images of Black Male Sexuality," Isaac Julien and Kobena Mercer make the important point that "the call by gay activists to reject the heterosexist norms of the nuclear family was totally ethnocentric as it ignored the fact that black lesbians and gay men *need* our families" (168, original emphasis). Although potentially "contradictory spaces," these families "offer us support and protection from the racism we experience on the street, at school, from the police, and from the state" (Julien and Mercer 168). Jamal's relationship with his family—specifically with Cookie and Lucious—perfectly demonstrates both the black gay man's need for his family and the contradiction that need creates. While denied Lucious' acceptance, Jamal enjoys the unconditional support and love of Cookie. It is the bond between the two that initially helps to sustain Jamal when his father's acceptance is denied. Sprinkled throughout the season's earlier episodes are flashbacks of Cookie's love for and defense of Jamal— flashbacks which become increasingly absent as Jamal gets closer to his darker self.

On the one hand, these flashbacks demonstrate Cookie as a fierce protector and strong system of support for the son whom she knows is "different." The most poignant, and most notable of these flashbacks shows Cookie pulling a five-year-old Jamal from the trash bin in which Lucious had dumped him after Jamal sashayed into the living room dressed in his mother's high heels and make-up ("Pilot"). On the other hand, these flashbacks position Jamal in opposition to his father and his father's masculinity. While each of the flashbacks demonstrates Cookie's support and love, that love and support comes in spite of, and as a response to Lucious' resistance to the man he sees Jamal becoming. However, to say that Lucious is inherently homophobic might oversimplify his relation to his son. After all, even as an adult, Jamal continues to receive financial support from his father, as well as access to the company and its resources. In fact, Lucious includes Jamal alongside his heterosexual brothers, Hakeem and Andre, in his plot to find an heir for Empire Entertainment. Moreover, Lucious seems intent on ensuring that Jamal—and his brothers—do not have to suffer the racism that he encountered. Instead, Lucious' anxiety over Jamal's homosexuality seems to stem from both his notion of masculinity, which has been shaped by his need to navigate the street life of New York's drug world, and his desire to usher his company into the mainstream American economy.

Rather than simply labeling Lucious' actions as products of an inherent homophobia, that is, simply as products of his fear or resentment of Jamal's homosexuality, per se, I suggest understanding them as the products of an inability to balance Jamal's attempt at a politics of innocence against Lucious' own notions of manhood and American society. Because this innocence goes against the very foundation upon which Lucious has built his own "empire"— both his ethos as a virile, masculine man of the streets, *and* his record label— he is unable to accept Jamal, who he feels to be lacking. If, as Reid-Pharr asserts, black homophobia is an attempt to return a sense of innocence to an otherwise tainted black community, Jamal's homosexuality, then, becomes a scapegoat onto which Lucious can project his own fears and shortcomings. As Reid-Pharr states it, "to strike the homosexual, the scapegoat, the sign of chaos and crisis, is to return the community to normality, to create boundaries around blackness, rights that white men are obliged to recognize" (*Black Gay* 104). This is very much the case for Lucious' rejection of Jamal's homosexuality. When Lucious dumps Jamal into the trash bin, it is as he is attempting to build his rap career—as he is attempting to leave the poverty and "hustle" that he has known for years. And when Lucious tells Jamal that he should remain closeted, it is in the midst of Lucious' attempts to build his company into a publicly traded empire. Caught between his son's desire to live authentically as a gay man and his own desire to steer his company through the avenues of mainstream American consumerism, and the gender regulations that consumerism imposes, Lucious finds himself conflicted when it comes to his relationship with Jamal. Lucious built his empire through and upon the notion of a defiant and defensive, and yet passionate, masculinity. As he asserts, "where I'm from, [when] someone tries to test your heart, you take their head off" ("Unto the Breach"). Jamal's initial politics of innocence, I suggest, goes against this principle because it lacks the passion Lucious recognizes as masculine, and it is not until their musical duel that Lucious begins to see this passion expressed in Jamal.

Electricity: Towards Conclusions

In a rather ironic way, I began this essay on black and queer politics with a homage to Whitman, the father of American gay culture. But, honestly, seeing how this is an essay concerned with a critique of a politics of innocence, I can think of no better place to begin. Whitman not only circumvents contemporary notions of innocence—Whitman's innocence is not clean, but messy and perverse—but he also posits life as a constant flow of energy—as a constant state of flux. In this sense, there is an untapped source of power in his work. More importantly, contemporary enjoyment of Whitman is no longer a perverse (or as perverse an) action as it was in the late-nineteenth century. We openly find pleasure in watching the bad boy, Whitman flaunt his sexual energy, often embracing his energy as our own. Sadly, a similar pleasure is often not taken in contemporary queer characters. Ironically, contemporary queer characters are placed under similar pressures Whitman himself faced during his time: pressure to clean up his act. We have, in many ways, returned to a society run by a politics of innocence. As I have sought to demonstrate through my analysis of Jamal, innocence offers an unsustainable strategy for acceptance.

Notes

1. Henceforth, episode citations will include title only.
2. Here, I make a distinction between the flashbacks that serve to signify Jamal's queerness as a youth and his self-identification as a gay man. Although these flashbacks are important for pointing to Jamal's impending queer identity and for explaining the tension between him and his father, as retrospective glimpses, they do not carry the same political burden I associate with his adult gay identity.
3. Without veering off the course of the present argument, I must take a moment to acknowledge the powerful ambiguity of Jamal's statement—a part of which I gesture towards with the parenthetical insertion of "the" and the capitalization of "empire." Jamal's statement leaves one wondering just which empire he plans on taking: Lucious' throne as a revered musical presence? The company, Empire, itself? Head of the family legacy? As the series continues to unfold, it seems that all, and yet none of these could be possible answers to what he means.
4. While I only focus on the first season in this paper, I find it important to note that Jamal's fame and creativity increase exponentially in the second season, after he has learned to more effectively balance the demands of his father with his own desires. Still, as the show continues to develop, his ability to balance the two is a work in progress.

Works Cited

Asante, Molefi Kete. *Afrocentricity*. Trenton: Africa World Press, 1989. Print.
Beach, Christopher. "Walt Whitman, Literary Culture, and the Discourse of Distinction." *Walt Whitman Quarterly Review* 12.2 (1994): 73–85. Web. 14 April 2016.
Bunn, Curtis. "Five Reasons Why 'Empire' Is Winning TV Ratings but Failing Black People." *Atlanta Black Star*. Atlanta Black Star. 28 Jan. 2015. Web. 18 Oct. 2015.
Cohen, Cathy. "Death and Rebirth of a Movement: Queering Critical Ethnic Studies." *Social Justice* 37.4 (2011–2012): 126–32. Web. 31 July 2014.

Edelman, Lee. *No Future: Queer Theory and the Death Drive.* Durham: Duke University Press, 2004. Print.
Empire: The Complete First Season. 2015. Prods. Lee Daniels and Danny Strong. Twentieth Century Fox, 2015. DVD.
Fanon, Frantz. *Black Skin, White Masks.* New York: Grove Press, 1967. Print.
Ferguson, Roderick A. *Aberrations in Black: Toward a Queer of Color Critique.* Minneapolis: University of Minnesota Press, 2004. Print.
Freud, Sigmund. *Three Essays on the Theory of Sexuality.* Trans. James Strachey. New York: Avon Books, 1972. Print.
Hemphill, Essex. "Introduction." *Brother to Brother: New Writings by Black Gay Men.* Boston: Alyson Publications, 1991. xv–xxxi. Print.
Jagose, Annamarie. *Queer Theory: An Introduction.* New York: New York University Press, 1996. Print.
Julien, Isaac, and Kobena Mercer. "True Confessions: A Discourse on Images of Black Male Sexuality." *Brother to Brother: New Writings by Black Gay Men.* Ed. Essex Hemphill. Boston: Alyson Publications, 1991. 167–73. Print.
Killingsworth, M. Jimmie. "Whitman and the Gay American Ethos." *A Historical Guide to Walt Whitman.* Ed. David S. Reynolds. New York: Oxford University Press, 2000. 121–51. Print.
Puar, Jasbir K. *Terrorist Assemblages: Homonationalism in Queer Times.* Durham: Duke University Press, 2007. Print.
Reid-Pharr, Robert. *Black Gay Man: Essays.* New York: New York University Press, 2001. Print.
———. *Once You Go Black: Choice, Desire, and the Black American Intellectual.* New York: New York University Press, 2007. Print.
Walcott, Rinaldo. "Outside in Black Studies: Reading from a Queer Place in the Diaspora." *Black Queer Studies: A Critical Anthology.* Eds. E. Patrick Johnson and Mae G. Henderson. Durham: Duke University Press, 2005. 90–105. Print.
Warner, Michael. *The Trouble with Normal: Sex, Politics, and the Ethics of Queer Life.* Cambridge: Harvard University Press, 1999. Print.
Whitman, Walt. *Leaves of Grass: The Original 1855 Version.* Mineola: Dover Publications, 2007. Print.

Merlin
Magician, Man and Manipulator in Camelot

CAROLINE WOMACK

Over the centuries, scholars have attempted to figure out who the original Merlin was and when he lived. Unfortunately, he is most likely a conglomeration of various historical characters mixed with medieval magic and creativity. In the 2011 Starz television series *Camelot,* the producers strive to maintain a close connection to Malory's *Le Morte d'Arthur,* using it as their primary source. Starz presents an effective version of Merlin that is as close as a modern audience will get to a realistic version of an historical Merlin. In this treatment, Merlin is presented as a man with dangerous powers who manipulates others for the good of the idealistic society he hopes to create.

In Arthurian legend, Merlin is the archetypical wise old man, but this archetype is closely linked to that of the trickster. According to Jung, the archetype of the wise old man is a duplicitous character that is not entirely morally good. He states, "the figure of the wise old man can appear so plastically ... that it takes over the role of guru" (126). In this case, "plastically" can refer to how the wise old man can take many different forms, such as Merlin changing his own physical shape or changing the appearance of others. It can also mean that Merlin is insincere. This mirrors Hyde's study in his book *Trickster Makes This World: Mischief, Myth and Art.* Hyde argues that the "trickster is a mythic embodiment of ambiguity and ambivalence, doubleness and duplicity, contradiction and paradox" which inevitably help people learn to solve their own problems (7). In spite of this mutability, who would aid the protagonist into achieving his goals if not for the wise old man? Clearly, until Merlin's ultimate demise at the hands of a woman, he is the moral guide in Arthurian mythology driving Arthur's rise to power, but is he on the side of good or evil?

Due to the lack of historical records, no one is quite sure who Merlin truly was. *The New Arthurian Encyclopedia* claims that he was "a Briton who was crazed by a battle at Arfderydd in Cumbria, ca. 575. He wandered through the Caledonian Forest in Southern Scotland, an inspired madman, gifted or cursed with second sight" (Lacy 319). In Layamon's *Brut*, Merlin is the son of a fifteen-year-old princess. Like in the Ancient Greek story of Danae, a knight in gold comes to her in her dreams and, from this visitation, Merlin is conceived (Layamon). Merlin's father has been called a devil, an elf, or a god, which is similar to the Merlin story from Geoffrey of Monmouth's *The History of the Kings of Britain* where Monmouth says that Merlin is the son of an incubus and a nun. This explains Merlin's affinity for magic and connects him to Christian theology as well (Archibald and Putter 203).

Malory gives the reader next to nothing about Merlin's background in *Le Morte d'Arthur*, and *Camelot* accordingly follows suit with Merlin being a private and secretive individual. Very little is offered about Merlin's origin. He is intelligent, has the ability to see the past and future, and can "do things others believe impossible." He was not born with magical ability, nor did he study it. He says that he is "form made manifest" and that there are forces in the fire, wind, water, and earth, which Merlin has the ability to shape to his will. When Arthur asks Merlin about the story of his birth, Merlin insists that the "details aren't important." He wants Arthur to "stop asking about me and think about who you are" ("Homecoming"). Identity is key in making Arthur into the king that people will want to follow. It can be argued that Merlin's past was not as pure or noble as the future that he intends for Arthur and for Camelot; however, in not revealing his past or his past mistakes, he can take on a persona of a wise old mentor without physically appearing as an elderly man. Merlin also makes a point of not forming intimate relationships with other people, but makes sure to get close enough that he is able to earn their trust and respect. At one point, he claims that people who get close to him get burned ("Justice"). He later informs Igraine that he ran away for a long time before aiding in Arthur's conception and serving in Uther's court ("Igraine"). It is not known how long Merlin remained at Uther's court, whether he lived there or came and went as he pleased. It is only briefly mentioned that Merlin was one of Uther's advisors. What is most surprising is that this version of Merlin is reluctant to do any magic whatsoever. This essay examines the character of Merlin in three facets: Good Magician, Bad Manipulator, and In-Between Man. The section The Good Magician will disseminate the concepts of magic and how it is presented in both Starz's *Camelot* and its source texts. Though magic stems from something otherworldly and dark, as a magician, Merlin struggles to avoid using it and thus is a Good Magician. The In-Between Man examines Merlin as a human being to fully connect how he is presented in *Camelot*, exploring the historical character as a fallible man rather than a demon-sired, supernatural wizard. Finally, The Bad Manipulator will focus on Merlin's political machinations and its consequences as he, through Machiavellian strategies, aids in gaining the throne for Arthur.

The Good Magician

> There's a certain choice of the look and a certain choice of the way he wields his magic.... We wanted the magic to be something very organic, elemental, true to [Merlin as] a pagan character.
> —Fiennes, *AssignmentX*

The producers of *Camelot* attempt to create a realistic version of Arthurian legend in order to display how the legend surrounding Arthur's reign might have arisen. In an interview with *Collider*, *Camelot*'s show runner Chris Chibnall states, "We are telling the story of Camelot, using Thomas Malory's *Le Morte d' Arthur*.... We're starting right from the birth of Arthur, and we'll go through and try to tell the truth that lies behind the myth." Even in the Starz interpretation, in spite of the desire to make it more "realistic," magic is still a huge component in the man behind the myth. There are several occurrences in Malory where Merlin uses magic to disguise himself, to disguise others, to share prophetic visions, and to instruct Arthur; however, in *Camelot*, Merlin is a reluctant wizard and a wise man that has learned it is better for things to come about naturally than to mess with the addictive, somewhat dark, and primitive *magic*. According to Christopher Dean, "Malory plays down Merlin's magical powers, preferring instead to present him as God's agent here on earth" (63). This fits in well with *Camelot*'s vision, attempting to remove the magic from the legend or at the very least sideline its importance.

Just as some consider Merlin's magic to be from the devil[1] in Malory and Laymon's *Brut*, *Camelot* maintains that magic is less of a gift than it is a curse. Though people can learn to wield the innate magic within them, allowing such an action is more like an addictive drug than something beneficial to the world. Several times in the TV series, Merlin says that using magic comes with a cost, but the cost is never something a person can anticipate. He says that magic costs "the body, the soul, and everyone around [him]" ("Three Journeys"). He warns Morgan that he is "stronger to choose not to use magic" ("Guinevere"). Showing the limitation of his abilities, Merlin has the ability to heal, but refuses to aid a dying man, telling the others in his company that he does not have the power to stop people from dying if it is their time, nor can he bring people back from the dead ("Three Journeys"). In this same episode, Merlin uses his magic to set some leaves on fire to distract Leontes from having his dislocated shoulder fixed. Gawain suggests that Merlin can learn to control his magic if he wanted. After this, there is a close-up of Merlin, which suggests that he would consider the idea, but nothing further comes of this musing.

In addition to the innate ability to control the earthly elements, Igraine observes that Merlin "hasn't aged a day" since she last saw him which was around twenty years ago. Either he is immortal, or ages more slowly than other people. Nevertheless, in his long life, it is clear that Merlin has become wise. Because he appears as a middle-aged man, it gives him physical strength or prowess that would not be apparent in an elderly man. The choice of the directors to make him appear younger prevents Merlin, as the man, from needing the magic that sets him apart from other people and places the dichotomy of good vs. evil in immediate conflict with each other rather than placing magic at the forefront of the moral battle.[2]

The de-emphasis of Merlin's backstory and demonic father in *Le Morte d'Arthur* merely adds to the air of mystery surrounding this man who does not seem to age. Campbell, in her examination of Robert de Boron's *Merlin*, argues that there is a direct correlation between demons and morality. She attests "the demonic seems to externalize man's own capacity for moral failure, the psychological dimension of Robert's devil representing the internal conflict of an individual at odds with his/her conscience" which is applicable to Merlin as the son of the devil (48). Merlin could be seen as a manifestation of the guilt of his mother's insatiable desire or the symbol of the struggle between good and evil within all mankind.

Even when *Camelot*'s Merlin intentionally uses the power in order to aid Arthur, there is always a cost. One instance of the cost of magic is the death of Excalibur, daughter of the sword maker Caliburn. Merlin saw a vision of Caliburn using the new sword to assassinate Arthur. For this reason, Merlin attempts to steal the sword to prevent the sword maker from returning to Camelot with him. In this attempt, Caliburn nearly overpowers Merlin. Merlin then kills Caliburn in a magically enhanced blaze of fire. Excalibur, Caliburn's daughter, then steals the sword intended for Arthur. Merlin pursues Excalibur to a lake and Merlin freezes the water, unintentionally trapping the girl underneath the ice. As she is drowning, she thrusts the sword through the ice and Merlin takes it. He is unable to save the girl. Merlin did not want or intend the sacrifice, but he brings the sword to Arthur ("Lady of the Lake"). No matter the intent, Merlin cannot fully control his powers and the unintentional death of an innocent girl furthers the theme that every action has consequences. This is why Merlin, for the most part, does everything he can to avoid using magic.

Another harsh consequence of using magic in *Camelot* is the physical reactions such as coughing up blood, bleeding from the eyes, and becoming severely ill. While Morgan freely and carelessly uses her magic, indifferent of the physical consequences to her body, Merlin avoids it. Merlin does not want to cause pain through what he considers to be dark or evil magic. Merlin told Arthur's knights that he hates the power inside him and every hour is a struggle to avoid using it. He says the power pulls him "to the dark" ("Three Journeys"). Magic, Merlin says, is like a second emotion that he struggles to ignore, but there are times when the struggle overcomes him ("Lady of the Lake"). He is like a drug addict after rehab, struggling not to use the magic that always ends up hurting the ones he loves.

Most of the time in the TV series Merlin's attempts to change fated events comes too late, as in Igraine's death. Although Merlin refuses to use magic the entire time Morgan holds Igraine and him captive, he attempts to heal a mortally wounded Igraine. Knowing the limits of his abilities, he would have ended up killing himself to save her, had she not insisted that he "protect Arthur." Rather than follow through on her request, an ultimately broken Merlin ends up leaving at the end of the series finale. He has already lost too many loved ones and cannot stay around to see other people he cares about die.

In Malory, on the other hand, the only "curse" of having magical abilities is that Merlin is unable to stop things from happening. In spite of all of his prophecies and warnings, Mordred is conceived, Lancelot and Guinevere fall in love, the sword and scabbard are stolen by a woman Arthur trusts, the Round Table disbands, Arthur's kingdom collapses, and Merlin is trapped under a stone. It is as though fate will always have its way, no matter what Merlin tries to do to stop it. It could also be argued that Merlin's own weakness, his infatuation with a woman, inevitably leads to his downfall. Either the magician's downfall is caused by his deep connection to humanity, the human emotion of love that is not rooted in the spiritual world, or this is the moral tale of a trickster who ends up being the one who is tricked.

The In-Between Man

> I wanted to get away from that iconic look of a wizard with the pointy hat and staff and long beard.
> —Fiennes, *Digital Spy*

In the media over the years, there have been several renditions of Merlin, ranging from comical to menacing. There is a bearded advisor (*Knights of the Round Table*, 1953), a former mentor of the king (*Camelot* musical, 1967), an enigmatic wizard (*Excalibur*, 1981), an aloof and at times confused wizard who ends up telling stories for money (*Merlin* TV mini-series, 1998), a Druid bard from Avalon (*Mists of Avalon*, 2001), an old Woad in the woods (*King Arthur*, 2004), a middle-aged trickster (*The Last Legion*, 2007), and a teenaged servant just learning to use his powers (*Merlin* TV series, 2008). Motivation, emotions, and personality are not explicit in Malory, whereas in any dramatization of the Matter of Britain, the audience has a better idea about the characters' motivations. Every scene in Malory is stated straightforwardly as to who said or did what, rather than *why* a character said or did something. It is up to the reader to draw conclusions from dialogue and action, rather than through any character development by the author. Emotions are both seen and heard in a movie or TV series, allowing viewers to connect with the characters and better understand their motives which is unavailable in Malory. This is essential to humanizing the man behind the myth, creating a more palpable Merlin.

In literature, Merlin's life, though somewhat of a mystery, is not nearly as long as portrayed in other versions of the tale. In an essay from *The Book of Merlin: Insights from the Merlin Conference*, Geoffrey Ashe attests that an historical Merlin was most likely in his forties when he left Arthur for the cave. He was a child when he foretold Vortigern's doom. Goodrich also maintains that Merlin would have been about twenty-five years older than Arthur (333). So why is it that Merlin is typecast as an elderly man?

From Tiresias to Dumbledore, archetypical mentor guides are often seen as elderly men with long grey or white beards, carrying a staff, and imparting their knowledge. While Tieresias, Dumbledore, and Gandalf are in fact quite old by the time they meet their protégés, Obi-Wan and Merlin are hardly old. Nor are Obi-Wan or Merlin "old" when they leave their protégés. Obi-Wan was fifty-seven when he died and Merlin was most likely in his forties when Nimue trapped him. They were not "old" by modern standards, but perhaps because they are wise they must be viewed as old, indicating wisdom is only gained through age? Perhaps it is because people began to live longer and they needed to cast the wise man as an elderly man. Or perhaps it is Nimue's seduction of Merlin, causing his impotence that created this image of an older man with a long beard.

Malory's Merlin is a complex and contradictive character. He wants Arthur to prove his worth, but then is angry about the bloodshed. He is playful and teasing, yet instructive and incredibly serious. The only time Malory truly shows him as human is in the way Merlin desires a woman that he knows will be his undoing. He knows she will lead him to his destruction, yet he tells her everything she wants to know. This is the only fallibility that we see from Malory's version of Merlin and the only clue as to Merlin's human nature. The rest of the time, Merlin is aiding kings in the creation of the

Round Table, prophesying events of the future, or playing tricks in a way to offer Arthur some sort of moral guidance (Batt 51).

While *Camelot* presents a Merlin that gives moral advice, he cannot disappear at will or save people from death. At first Merlin resists all temptations of the flesh. Although his desire for Igraine and their intimate relationship is strange, because Igraine seems too quick to forgive Merlin for arranging Uther to rape her and consequently stealing her child, this shows Merlin's earthly desires for both physical and emotional connections to other people. Later, when Morgan stabs Igraine, Merlin attempts to heal her and bleeds from his eyes in the process. His ensuing grief over her death prevents him from staying in Camelot at the end of the series.

Camelot's Merlin is also humanized through the emotions he experiences in the episode with Caliburn and Excalibur. In his anger, he kills the man, but does not regret the man's death, for it was in protection of Arthur. However, he experiences such a deep regret over the accidental death of Excalibur, agonizing over his use of magic that he starts a fight in a pub in order to punish himself. He wants to feel physical punishment for what he considers to be murder and struggles with his guilt. In addition to suffering from the self-inflicted wounds, he also experiences withdrawal for having used magic. For several days, the withdrawal prevents him from eating and sleeping as much as he normally does. Merlin's own struggle with public honor and private shame is a Malorian dichotomy. On countless occasions, the knights, and even Arthur himself would rather die than live with shame. In Arthur's confrontation with Accolon, he attests, "I had liever to die with honour than to live with shame" (Malory 66). Proper knights, even if it cost them their lives, would not admit they were recreants, liars, or had committed a deed which they had not committed according to their conceptions of honor.

Another aspect of a more humanized Merlin in the TV series is his lack of faith in the new Christian religion. While he is portrayed as a pagan, he does not embrace any religion. In a conversation with Leontes, who is a Christian, Merlin unintentionally insults him, saying that there are no gods. Leontes requests that Merlin not mock his belief in God, and then Leontes insists that Merlin's magic comes from God, even calling Merlin an angel. After laughing uproariously, Merlin replies, "I've been called many things, but not that" ("Three Journeys"). He knows he is just as human as Arthur's knights are, though it is interesting that while Merlin is an atheist, he learns enough of Christian religious beliefs to be able to use it to his advantage later in manipulating Leontes.

In Malory, Merlin is absent for a good stretch of time during Arthur's reign because Malory wanted to maintain the Christian themes. He could not have a pagan or a druid instructing the otherwise Christian king. Oftentimes, Malory has Merlin reference God in order to persuade Arthur to do the right thing. For example, when Arthur smites kings in North Wales, proving that he is a battle-worthy king, Merlin rides in on a black horse to stop Arthur from further slaughter. "Thou hast never done! Hast thou not done enough? of three score thousand this day hast thou left alive but fifteen thousand, and it is time to say Ho! For God is wroth with thee, that thou wilt never have done..." (Malory 31). It is Merlin, speaking on behalf of the divine, who is angry with Arthur. Merlin uses the name of God to increase Arthur's guilt over the decisions he made in battle. Similarly, in *Camelot*, the bloodshed disgusts Merlin. He wants the knights to fight by guile rather than by the sword ("Homecoming"). Merlin uses his own guile to persuade the knights that his way of doing things without a sword is the right way to win the hearts and minds of the people of Camelot. "We are in the fight of our lives for the soul of this country" ("The Sword and the Crown"). This can be taken as the Christian soul of the country, using this diction to persuade a Christian king to do the right thing, what is morally good according to Christian virtues, to motivate them in a positive way. Merlin clearly does not believe in any gods, but chooses the word "soul" specifically to address the Christian knights.

The Bad Manipulator

> The great thing about Camelot that Chris has reinvented and adapted is that the magic really lies in the political essence of the piece. Yes, there will be some dark arts, and we'll see people changing shape and things disappearing, but over and above all of that, the real potent magic is the birth of the legend.
> —Fiennes, *Collider*

From the very first chapter of *Le Morte d'Arthur*, Merlin is shown as an intelligent, cunning, and politically perceptive man who Sir Ulfius sought out to aid the king. Merlin orchestrates Arthur's conception; he tells Ector how to instruct the boy as Arthur grows up; he has Uther sign his kingdom over to Arthur on his deathbed and Merlin brings Arthur to Camelot. Everything Merlin does has a higher purpose, but whose purpose does it serve? Is he merely serving kings, or is he serving himself? When Merlin changes Uther's appearance so that he may sleep with Igraine, he is not merely doing Uther a favor. Merlin is bringing about Arthur's prophetic birth to save the people of Britain (Stewart 43).

Just as in Malory, *Camelot*'s Merlin took Arthur away from his parents as a child to be reared in Sir Ector's house. The choice seems deliberate for not only does a good man raise Arthur, but he also makes sure that Arthur is literate in the art of war. Merlin specifically told Ector how to raise Arthur before he left. The books from Ector's house are considered more valuable than gold to Merlin and later on, they make a journey to recover the books to start a library in Camelot. Knowledge is important in molding a king. The library gives the new king a "place to think" and hopefully a place to learn from the atrocities of the past in order to bring about a new way of life for the people of Britain ("Three Journeys"). Merlin does not arrive in time to prevent Uther from dying which leads to the conclusion that he did not intend to save Uther's life. He was, however, just in time to have Uther sign a legal document indicating that Arthur was his son and rightful heir. This is questionable behavior on the part of Merlin for three reasons: Merlin didn't bother trying to save Uther's life; he didn't give the dying man the chance to object to signing the document, and since Uther died before finishing the signature, Merlin signed the rest of it himself. Forging a legal document in attempt to sign over the throne to Arthur is deceitful behavior, even though it was for, as Merlin views it, a good reason. At one point Merlin refers to Uther as a "barbarian" and in the first episode, Merlin has visions that are most likely flashbacks from Uther's reign that are filled with bloodshed.

Merlin believes wholeheartedly in his vision of the future free of bloodshed. In the first episode, Merlin sells Arthur a great idea: "This realm needs a leader, better than anyone that's come before.... We are going to build a land full of hope and honor, where fear is extinguished, to which people will flock from far and wide, seeking out our beacon of light" ("Homecoming"). Hope and honor are ambiguous terms, but persuasive to a young man that still is innocent enough to have hope in a better future for all of Britain. Arthur, in this case, becomes the "beacon of light," the symbol of hope in a nation ravaged by war. Merlin insists that Arthur is the key because he has seen a vision of the future. He has seen the "darkness of man" and does not want it to happen again. "The past," he says, "doesn't matter. Define yourself in the present and you might rule in the future" ("Homecoming"). The knights who served with Uther also agree with Merlin's vision of the future and are quick to stand by Arthur. As in Malory, according to Armstrong, Merlin's true purpose in the narrative is to instill Arthur on the throne and set up a secular form of chivalry through the Round Table (27).

Creating a mythology behind Arthur is essential to obtain power over the people in the war-weary land. In both Malory and *Camelot*, Merlin devises a test to prove to people that Arthur is the true king of England. It is implied that *Camelot*'s Merlin created the "Sword of Mars" myth and placed the sword in the waterfall many years ago in anticipation of Arthur's coming. Merlin says that "a king exists primarily as an idea. We [must] persuade the people to believe in the idea of [Arthur]" ("The Sword and the Crown"). The sword in the stone episode is altered so that it is not merely an event for knights to witness, but for a whole crowd of common people. Merlin's reasoning for this is that he wants as many people as possible to witness the feat so that in a bastardization of John 1.1, "In the beginning there was the word: and the word spread: and the people came." Like Jesus, Arthur has a miraculous birth, and it is his *destiny* to save Britain. Most movie and TV adaptations show Arthur surrounded by knights who automatically kneel in reverence to their new king. In this case, they want to illustrate the fact that Arthur is also the "people's" king. Additionally, rather than have him pull it out of an anvil seven times as in Malory, it is taken from a treacherous waterfall. This test not only aids in Arthur gaining courage, but also creates the myth of Arthur's divine right of kingship.

The first mythical Sword of Mars is broken and *Camelot*'s Merlin is in need of another enchanted sword to replace it. Merlin uses the mythology of the "Lady of the Lake" to cover up two murders. Instead of a woman with mystical qualities that gives Arthur the sword, she is merely a daughter of the blacksmith who created Arthur's second sword. Rather than tell people the truth about Caliburn and Excalibur's deaths and the inevitable *cost* of Merlin using magic, he creates a wonderful story that increases the people's view that Arthur is king by divine right, both Roman and Pagan. Merlin recalls:

> Out of the mist was a woman who called to me like a siren. From within the lake, she stretched out her arms with this sword, clutching it. I took the sword and thanked her. She smiled and slipped back into the water. And as she did, she said: "This is the sword of King Arthur. This is Excalibur" ["The Lady of the Lake"].

Referencing a mythological siren and using vivid imagery of the Lady of the Lake affixes Arthur's reign to that of Ancient Greek legends. Having the mythical woman give Arthur the sword shows that the gods endorse Arthur's reign. It is part of actantial model which is so often found in folktales (Hébert 71). Rather than magically installing Arthur as a king, he wants Arthur to earn it and to earn the trust of the people. Merlin is the spark that made Arthur, but Arthur has to become his own man by proving himself to the people.

The supernatural has many functions in Malory such as aiding or serving as moral challenges to knights on their quests. Similar to Merlin's myth creation in *Camelot*, in Malory "the variousness and ambiguity of the supernatural and its practitioners became powerful tools in the creation of drama and suspense" (Archibald and Putter 207). When Merlin and Arthur arrive at the legendary lake, Merlin declares, "within that lake there is a great rock, and therein is as fair a palace as any on earth, and richly beseen. And this damosel will come to you anon; and then speak ye fair to her that she may give you that sword" (29). The Lady of the Lake is not given any other introduction. Clothed in "white samite" and requesting that Arthur owe her a favor "when I see my time" adds an air of mystery as well, adding a mystical assurance and acceptance of his divine right of kingship.

Although Malory's Merlin helps instill Arthur on the throne, he also encourages infanticide after revealing Arthur conceived a child through incest. The section begins:

> Then King Arthur let send for all the children [...] for Merlin told King Arthur that he should destroy him and all the land should be born on May-day. Wherefore he sent for them all, on pain of death; and so [...] all were put in a ship to the sea, and some were four weeks old, and some less [31].

Merlin insists on this immoral act and Arthur, in turn, follows through with it. Even though Arthur orders the death of hundreds of infants, Merlin is equally accountable. The final episode of *Camelot* shows Morgan sleeping with Arthur, using magic to make him think that she is Guinevere. It can only be conjectured that, had the series continued, Merlin would make the same suggestion to Arthur: to have all the four-week-old infants sent off in a ship to drown at sea. Although the crime, in the case of *Camelot*, originates with Morgan's intentional plans to destroy her brother, the unborn child is the true victim.

Certain people, namely women, interfere with Merlin's plans in both *Le Morte d'Arthur* and *Camelot*. In Malory, Arthur wants to marry Guinevere, but Merlin warns him not to. He tells Arthur that Lancelot and she will fall in love and that it would be dangerous, but Arthur ignores Merlin's advice to satisfy his own desires (50–51). Similarly, in *Camelot*, Arthur wants Guinevere, but she is married to someone else. "Everything you are right now is because of me," Merlin angrily tells Arthur, wanting him to listen to his advice ("Guinevere"). He reminds Arthur that a king's duty is to his people first to help Arthur become less selfish. He also does not want women jeopardizing Arthur's rule.

Inevitably, *Camelot*'s Arthur refuses to listen to Merlin's advice and sleeps with Guinevere before her wedding which leads to Leontes discovering the truth. In his rage, Leontes nearly kills Arthur, but Merlin stops him, saying he will deal with Arthur, preventing Leontes from harming the king. Merlin then tells Arthur to cover up the affair because the needs of the realm come before personal relationships. Merlin also convinces Leontes to fight for Arthur one last time, to defend Bardon Pass. Using specifically chosen diction, Merlin appeals to Leontes' honor, saying, "Fulfill your duty to your king or your honor counts for nothing" ("Reckoning"). The chivalric duty to his lord prevents Leontes from leaving Camelot and Merlin succeeds in persuading him to fight at Bardon Pass which inevitably leads to Leontes's death. Merlin orchestrates Leontes's death to cover up Guinevere's infidelity so that Arthur's kingship would remain spotless. In Malory, honor is earned through military prowess, but is also maintained by public perception. If a knight were to lose face by dishonorable actions, such as refusing the king's request to fight in a battle, he would have to contend with public shame.³ In this case, Merlin covers up the affair so that Arthur remains flawless in the eyes of his fellow knights. As his knights prepare to depart, *Camelot's* Merlin calls out to "be victorious for Camelot" ("Reckoning"), a subtle reminder of the importance of fulfilling their duty to the king.

Conclusion

Camelot is a worthy adaptation of Malory in its use of Merlin's character even though he does not wish to use his magic. In this retelling of the classic tale, it makes sense that Merlin wants Arthur to be king in his own right. What Merlin steadfastly believes in is Arthur and by the end of the series, Arthur finally has faith in himself by accepting his destiny. Merlin has also fulfilled his role as Arthur's mentor. He has made sure that Arthur received a good education, invented the myth surrounding Arthur's kingship, and installed him on the throne. Unfortunately, Merlin does not burden Arthur with his prophecies of the future as he does in Malory. He does not tell Arthur about Lancelot and Guinevere or that a woman Arthur trusts will steal the sword and scabbard. Instead, he asks Arthur to let him go and insists that Arthur is stronger than he is. Arthur asks Merlin if he will return and Merlin's response is, "You will be great" ("Reckoning"). Though it is not the answer that Arthur wants to hear, it is sufficient and as Arthur's gaze returns to the land below the rampart looking out at his kingdom, it gives a sense of finality of their meeting as well as the hope for the future of Arthur's reign.

It is problematic to label Merlin as good or evil when the text operates on a mythic level. As a man in the Starz series, he is not entirely morally good because in spite of his powers he still allows bad things to happen. He is also not wholly evil because he truly wants the best for Britain and believes that the choices he makes are for the benefit of the "greater good." "Camelot," Merlin insists, "isn't built on magic, but on people, on their faith" which is something that Morgan fails to realize by abusing magic, using fear to manipulate people, and dealing underhandedly with her supporters to attempt to gain control over the land ("Reckoning"). It is the people of Britain, not the supernatural that are in charge of creating their own destinies by making choices. He knows Arthur is the key to saving Britain, but he also knows if he uses sorcery, the people would be against Arthur. For this reason, there are many instances when Merlin refuses to use magic, such as when Morgan holds him captive. Merlin believes that Arthur's misplaced loyalty to his sister will be revealed more readily this way instead of using magic to free himself.

In truth, magical power is a symbol for political power in the series. If Merlin had chosen to operate as some sort of god, as Morgan endeavors to, he would not be allowing the concept of free will. Fiennes states, "He doesn't want to taint [Arthur's kingship] with the arts that he's involved in. He wants this young boy to be that seed of power, to become King Arthur through absolute pure integrity" (*AssignmentX*). In allowing bad to happen instead of relying on his magical powers, he also allows the freedom of choice. As in every great tragedy, as in the story of Arthur's kingdom, the choices the protagonist makes inevitably leads to his peripeteia.

Unfortunately, Starz did not order a second season, so viewers will not know just how Merlin's character would have developed in the series. Merlin's departure in literature has had several moral interpretations over the years. It been interpreted as "an image for the taming of unbridled power of the subconscious" (Taylor and Brewer 13), symbolic of the human drive for power which inevitably leads to that destruction, the beginning of the fall of the Round Table "for its human weakness and human imperfection that ultimately bring to ruin the noble chivalric ideal" (Dean 63), the abandonment of God since now "the Arthurian world is left to govern itself without much external influence" (Radulescu 130), and finally symbolic of the fact that magic is dangerous and must be contained since "Arthur and his knights must prove themselves without constant advice and help from magicians" (Archibald and Putter 14). However, it could have also been seen as the removal of pagan ways from an otherwise Christian narrative. A true Christian king could not have a pagan or atheist advisor. Perhaps, in *Camelot* this is Merlin's final departure as it is in Malory, leaving Arthur early on in his reign. It could be that Merlin is taking the initiative now because he realizes that he must leave Arthur to govern on his own and give more credibility to his reign. In Malory, Merlin is only mentioned a few times in passing and not remembered at the end of the grail quest or even at the Battle of Camlann. Thankfully, we have T. H. White to thank for Merlin's dream vision to Arthur the night before the battle to reconcile the lack of commemoration of an important figure in Arthurian legend. In doing what he can to attempt to save Britain, is Merlin a hero or ultimately good? In *Camelot*, he is a visceral man with desires of the flesh, the need to eat and sleep, and stands apart with his unusual, innate powers. He is a flawed man who strives to do good in spite of his murky past. Starz achieves its purpose in showing Merlin as an historical figure by firmly rooting Arthurian legend in a pseudohistorical context in order to create a realistic version of the myth. Though *Camelot* contains ingenious recreations of Arthurian legend, in each of the aforementioned facets of Merlin's character, Starz is able to maintain his authenticity as a man with the potential for either good or evil.

Notes

1. "Beware...of Merlin, for he knoweth all things by the devil's craft" (Malory 99).
2. This is in contrast to most modern interpretations of Merlin. For example, in *That Hideous Strength* by C. S. Lewis and *The Dark Is Rising* by Susan Cooper, magic is at the forefront of the battle between Good and Evil, Light and Dark. "Because modern writers tend to make Merlin a man of power and wisdom ... he often fights as a champion of righteousness" (Dean 67).
3. For example, before Sir Blamor's confrontation with Sir Tristram, his brother Sir Bleoberis demands his brother die rather than shame his family. Sir Blamor swears, "Well may he smite me own with his great might of chivalry, but rather shall he slay me than I shall yield me recreant" (Malory 192).

Works Cited

Archibald, Elizabeth, and Ad Putter, eds. *The Cambridge Companion to the Arthurian Legend*. 1st ed. Cambridge: Cambridge University Press, 2009. Cambridge Companions Online. Web. 27 March 2016.

Armstrong, Dorsey. *Gender and the Chivalric Community of Malory's Morte d'Arthur*. Gainesville: University Press of Florida, 2003. Print.

Batt, Catherine. *Malory's Morte Arthur: Remaking Arthurian Tradition*. New York: Palgrave, 2002. Print.

Camelot: Season 1. Prod. Graham King and Michael Hirst. Perf. Jamie Campbell Bower, Joseph Fiennes, and Eva Green. Starz Media, 2011. DVD.

Campbell, Laura J. "The Devil's in the Detail: Translating Merlin's Father from the Merlin en Prose in Paulino Pieri's Storia di Merlino." *Arthuriana International Arthurian Society—North American Branch* 23.2 (2013): 35–51. Print.

Dean, Christopher. "The Many Face of Merlin in Modern Fiction." *Arthurian Interpretations* 3.1 (1988): 61–78. Web. 16 March. 2016.

Fiennes, Joseph, and Chris Chibnall. Interview by Christina Radish. "Joseph Fiennes and Writer/Showrunner Chris Chibnall Interview Camelot." *Collider*. Web. 3 July 2015.

———. Interview by Abbie Berstein. "Exclusive Interview: Camelot star Joseph Fiennes Finds the Magic in Merlin." *AssignmentX*. Web. 3 July 2015.

———. Interview by Catriona Wightman. "Joseph Fiennes: 'Merlin like Obi-Wan, Rumsfeld.'" *Digital Spy*. Web. 3 July 2015.

Geoffrey, of Monmouth. *The History of the Kings of Britain*. London: Penguin Books, 1966. Print.

Goodrich, Norma Lorre. *Merlin*. New York: Franklin Watts, 1987. Print.

Hébert, Louis. *Tools for Text and Image Analysis: An Introduction to Applied Semiotics*. Trans. Julie Tabler. Version 3, 2011. Signo. Web. 8 Nov. 2015.

Hyde, Lewis. *Trickster Makes This World: Mischief, Myth and Art*. New York: Farrar, Straus and Giroux, 2010. Print.

Jung, Carl. *The Essential Jung*. Ed. Anthony Storr. Princeton: Princeton University Press, 1983. Print.

Lacy, Norris J. *The New Arthurian Encyclopedia*. New York: Garland Publishing, 1996. Print.

Layamon. *Layamon's Brut (British Museum Ms. Cotton Caligula A.IX)* London, 1993. Web. 16 Nov. 2015.

Malory, Sir Thomas. *Le Morte Darthur*. New York: The Modern Library, 1994. Print.

Martin, Thomas L. "Merlin, Magic, and the Meta-Fantastic: The Matter of *That Hideous Strength*" *Arthuriana International Arthurian Society—North American Branch* 21.1 (2011): 66–84. Print.

Radulesca, Raluca L., "'Oute of Mesure': Violence and Knighthood in Malory's *Morte Darthur*." *Re-viewing Le Morte Darthur: Texts and Contexts, Characters and Themes*. Eds. K.S. Whetter and Raluca L. Radulescu. Cambridge: D.S. Brewer, 2005. Print.

Stewart, R.J., ed., *The Book of Merlin: Insights from the Merlin Conference*. Ed. R.J.S. London: Blanford Press, 1987. Print.

Taylor, Beverly, and Elisabeth Brewer. *The Return of King Arthur: British and American Arthurian Literature Since 1900 [i.e. 1800]*. Cambridge: D.S. Brewer, 1983. Print.

"This isn't a democracy anymore"
The Walking Dead's *Rick Grimes* and the Post-Apocalyptic Good Cop/Bad Cop

ANNETTE SCHIMMELPFENNIG

> With solid pavement under your feet, surrounded by kind neighbours ready to cheer or to fall on you, stepping delicately between the butcher and the policeman, in the holy terror of scandal and gallows and lunatic asylums—how can you imagine what particular region of the first ages a man's untrammelled feet may take him into by the way of solitude—utter solitude without a policeman—[...]—utter silence. Where no warning voice of a kind neighbor can be heard whispering of public opinion?
> —Joseph Conrad, *Heart of Darkness & Other Stories* (1899)

Decaying houses, abandoned highways, rotten corpses, and marauding gangs in the streets. Post-apocalyptic films and TV shows present us with pictures that all have one characteristic in common, namely the significant lack of any kind of order. In AMC's *The Walking Dead*, one man constantly strives to reestablish what has been lost. Rick Grimes (Andrew Lincoln) is a loving father, a loyal friend and, most importantly, a sheriff. Even though there is no longer a legal system in the post-apocalyptic world, Rick still embodies his former profession as doing so likewise embodies his attitude to life. Such a character, a "moral compass," plays a significant role in the zombie narrative, which Bishop defines as follows:

> [The] classic zombie story—i.e., the apocalyptic invasion of our world by hordes of cannibalistic, contagious, and animated corpses—has remarkably specific

conventions that govern its plot and development. These generic protocols include not only the zombies themselves and the imminent threat of a violent death, but also a post-apocalyptic backdrop: the collapse of societal infra-structures, the resurgence of survivalist fantasies, and the fear of other surviving humans [Bishop 19].

Important to note here is the fear of other survivors. In the course of the now six seasons, Rick realizes that the main menace is not the undead, but the living.[1] To determine who poses a threat to the group of survivors, and his family in particular, and who might be a valuable addition to his crew, Rick heavily relies on the "good cop/bad cop-technique," a construct that generally involves an interrogation method based on conflicting behavior by one or two persons which is used in actual police operations. Rick (ab)uses the technique not only in interrogations, but he also assumes both roles for longer periods of time because, as will be argued in the following, doing so is Rick's tool of choice to establish his ideal hierarchy. This behavior, however, strongly depends on Rick's respective counterparts and their intentions. Rick adapts himself and his use of the technique to how he, and he only, assesses another character. Depending on their actions, Rick may first encounter them as a "good cop" and later on assume the "bad cop-role" and vice versa if he suspects them to threaten the well-being of the group. His demeanor as cop is therefore not static, neither consistently good nor bad, it oscillates between the two extremes. In the course of the show Rick's continuous decline of sanity, triggered by a series of personal losses, clouds his judgment and affects his reliability so that his use of the method becomes seemingly more and more arbitrary.

Thus, this essay examines the transformation of Rick Grimes, from a rational thinker to the borderline maniac in his personal "ricktatorship" and back. By comparing his character to the show's other leaders, the text will explore Rick's motives, be they altruistic or selfish, to demonstrate that his transformation is necessary to secure the group's survival. Ultimately, in a post-apocalyptic setting a character which has learned to alternate between good and bad is most likely to succeed in saving his own and other peoples' lives. This success, the show seems to argue, must be paid for by a loss of humanity.

"Who the hell are you?"—"Officer friendly"[2]

To better understand Rick's "natural leadership" (Sugg 804), it is necessary to analyze how Rick is introduced. He is the first character to appear which causes the viewer to bond with him; he functions as a guide through the post-apocalyptic setting. Rick is first shown wearing his uniform and driving a sheriff car through an alley of overturned cars. Without uttering a single word, his appearance thus establishes the position he assumes within the plot and, later on, the group as well. Rick can rely on the fact that his rank within society is constructed through the clothes he wears; it does not have to be established otherwise. The uniform speaks for itself, or in this case, for himself. In the early stages of the post-apocalyptic world the police as a social institution might not exist anymore, but the police man is still recognized as a reminder of it and the moral values it stood for. When he finally speaks his first words are "little girl, I'm a policeman" ("Days Gone Bye") to a little zombie girl he finds meandering around. Later on in the same episode, he tries to get gas and walks up to several houses shouting, "Hello? Police officer out here." Again, this suggests that Rick firmly believes that his title as a sheriff is enough to evoke feelings of trust and security for people willing to help him or in need of help themselves. Later on in the series, Rick uses his title for opposite effect, i.e., as a threat.[3] Rick is well aware of how to use the power his pre-apocalyptic position awards him, and when required, to embrace the role of the good or the bad cop. Both roles are generally defined as follows:

> "Good cop" [is] used to refer to roles conveying positive and supportive feelings such as warmth, friendliness, approval, respect, empathy, and sympathy to a target person; [...] "Bad cop" [is] used to refer to conveying negative and unsupportive emotions such as coldness, disapproval, lack of respect, and hostility [Rafaeli and Sutton 758].

While in many cop movies the roles are stagnant and attributed to one character each, it is also possible that a single character switches between them. In the beginning, however, Rick is the epitome of the good cop. Although he just awoke from a coma after being shot in the line of duty, Rick is immediately back at work. When Morgan (Lennie James) and his son Duane (Adrian Kali Jones) take the disoriented Rick into their home and provide him with food and clothes, Rick is thankful, but changes into another sheriff outfit, as if the uniform is necessary to rebuild the identity he lost while unconscious. Simultaneously, it is precisely this outfit which spares him from having to establish his position in a group of unknown people. Where most of the other survivors have to prove themselves to gain respect, it is simply assumed that Rick is in charge:

> Rick, the Sheriff turned group leader, is in charge because of implicit power dynamics, not simply his natural abilities. We see the drama that surrounds him coming not from a sincere will to fulfill his duty as a Sheriff to protect others, but instead from a desire for power and dominance over the group to protect his family [Berk 49].

Although back in Atlanta, Rick is saved by Glenn (Steven Yeun), he quickly switches from receiving Glenn's orders to ordering around the survivors he meets in the basement of a store. Good cop that he is, he feels the need to take charge and save them, thus instructs them to cover each other in the guts of a dead man, but not without checking his wallet for his name and photos. "He used to be like us" ("Guts"), he tells the others, as if to emphasize that as a cop it is his duty to remind the citizens of their humanity and save them from brutalizing each other. Interestingly, in the course of the show Rick will drastically change his good cop-attitude once he realizes that the sheriff outfit alone does not grant him and his family security, especially when he is confronted with people who benefit from the anarchy of societal collapse. Simultaneously, his formerly distinct insight into human nature fails him once he begins to rely on the authority of the bad cop routine.

Rick and Shane, Best Foes Forever

The first person to challenge Rick is, ironically, another cop, Shane (Jon Bernthal). Shane was initially Rick's best friend and partner, also present when Rick was shot. With Shane, a literal bad cop is introduced. Although both men are shown as typical buddy cops in the flashbacks, for the viewer Shane assumes the image of the enemy because he is first shown having intercourse with Rick's wife, Lori (Sarah Wayne Callies). Not aware that Rick was still alive, Lori engaged in a short-lived relationship with Shane, which she immediately broke off once Rick arrived at their camp. Shane, however, remains in love with her. As a result, "[t]he love conflict spurs the more violent male [Shane] to challenge his rival for group leadership" (Gavaler 15). Shane changes from being Rick's accomplice to his adversary. The fight for Lori's affection, which is never an actual fight as Lori decides to stay with Rick even though she is pregnant and unsure whose baby it is, seems to trigger a deeply-rooted rivalry between both men, which sends Shane into a downwards spiral. A first indication of this becomes apparent when Rick and Shane roam the woods around the camp. Shane secretly aims his gun at Rick, an incident which is observed by fellow survivor Dale (Jeffrey DeMunn), who only exclaims, "Jesus!" ("Wildfire"). With that action, Shane almost abuses the power, and the shooting skills, that come with his former job. When the group later arrives at the Centers for Disease Control and Prevention, they enjoy an alcohol-fueled evening together with the center's last surviving scientist, Dr. Jenner (Noah Emmerich). Shane interrupts the wanton mood by interrogating Jenner about the whereabouts of the other scientists. With Rick leading the way, the others heavily criticize Shane's behavior, even though it appears to be a legitimate question. Shane's frustration thus culminates in an attempt to rape Lori. It appears as if Shane has to assume the role of the maniac bad cop because with the return of the ultimate good cop Rick, the role he had is now taken; the worse Shane acts, the better Rick appears. This emphasizes how strongly the good cop/bad cop-technique relies on clear opposites. There is no gray area between Shane and Rick; it is impossible for one to be only slightly bad or temporarily good. Shane might start out as Rick's mirror, as in the pre-apocalyptic world they shared the same premise, namely to be white knights, but as Rick has come to occupy this role for himself, and the group only needs one leader, Shane is forced to rebel by becoming an antagonist because otherwise his loss of status within the group might be interpreted as weakness. Rick no longer complements him, and he is left alone. In a way this can be described as a bad cop-move by Rick against his friend. Rick is well aware that his fragile status as the good cop is built on Shane's refusal to provide comforting yet clearly false hope. He uses his friend as a negative example of a person who has lost faith, something the group craves as the morals start to decline, and thus isolates him.

In season two, Shane finds a temporary accomplice in Andrea (Laurie Holden) who feels dissociated from the group after the death of her sister and describes them as the "odd men out" ("What Lies Ahead"). While they share a short-lived fling, she is also subjected to his bad cop-attitude when he tells her "you shoot like a damn girl" ("Secrets"). Furthermore, he repeatedly mentions her dead sister to improve her aim. Andrea then decides to stay and the group thus remains together, mainly to find Sophia (Madison Lintz), Carol's (Melissa McBride) daughter, who is missing following a walker attack on the highway. This incident refuels the rivalry between Rick and Shane. Rick tried to rescue her by telling Sophia to return to the cars, but Sophia never made it back alive, as is revealed in the mid-season finale. Although her fate remains unclear until then, Rick repeatedly spurs the group not to lose hope while Shane casually mentions at various times that he thinks they will not find her alive. Again, Rick assumes the position of the good cop in his insistence on hope and his desire to provide comfort to those around him. Conversely, Shane, ever the bad cop, repeatedly confronts the group with hard truths and spurs conflict. In the end, however, this renders Shane the realistic one, but it also shows how little he thinks ahead. Rick knows that hope is the only thing that keeps the group united, which is also why he has to blight anyone who tries to destroy it. Dale senses the turmoil Shane causes and tells him openly "I know what kind of man you are" ("Secrets") but it remains open what exactly Dale means. A potential rapist? A murderer? A psychotic maniac? All of this applies and it is caused by envy. Shane envies Rick, his family and his position as the leader everyone blindly trusts, and this is what makes him unreliable. The post-apocalyptic world might desire the stability the good cop Rick provides but at the same time this is a utopian longing because it can only be achieved temporarily, as long as the good cop is in control. When Shane starts to challenge Rick, this introduces the slowly but steadily progressing leap of faith Rick experiences in the course of the show's seasons which forces him to become a bad cop himself.

In season two, the group finds a refuge at the farm of the vet Hershel Green (Scott Wilson) and his family. Hershel takes care of Rick's son Carl (Chandler Riggs), who had, accidentally, been shot by Otis (Pruitt Taylor Vince), Hershel's helping hand. In this part of the story it becomes clear that Rick is certainly capable of being a bad cop himself, which might also be one of the reasons for his long friendship with Shane. Although Shane remains the main source of trouble for the group, Rick increasingly besets the helping, but aloof, Hershel who wants the group to leave his farm once Carl has recovered. Rick considers the farm a safe haven and therefore does not want to move on. As Carl's condition stabilizes, he bad cops Hershel by telling him that Lori is pregnant and that it is either "a gift here" ("Pretty Much Dead Already"), or a death sentence out there, i.e., away from the farm. A dilemma is thus imposed on Hershel because can a doctor, even if he is "only" a vet, expel a pregnant woman from his farm without moral scruples? Shane may be ruthless, but it becomes clear that Rick also does not refrain from using morally questionable techniques to reach his goals. Only shortly after Rick proudly proclaims, "I'm not the good guy anymore" ("18 Miles Out"), which can be understood as a threat towards Shane and marks a turning point for Rick, Shane's further shenanigans confirm Rick's belief that there can only be one of them and a deadly confrontation appears inevitable. Especially because Shane not only questions Rick's leadership qualities, but also his ability to be a good father and husband. Shane finally dies twice, symbolically, once by the hands of his best friend, by being stabbed into the heart, and again by being shot by Carl, who follows into his father's footsteps. Like a divine intervention, the farm is immediately overrun by walkers and bursts into flames, robbing Rick once again of a potential home. The events on Hershel's farm ultimately drive Rick over the edge and the schizophrenic development of a split-cop personality is completed in the season finale. Regarding Shane's death he confesses, "I just wanted it over, I wanted him dead" ("Beside the Dying Fire") and he finally tells the group the secret Dr. Jenner told him and which he kept from them, namely that they are all infected with the fatal walker-virus.

At the same time, Rick experiences a new challenge. Before it was only Shane who questioned his leadership qualities. Now Carol voices what many of the survivors think: "We're not safe with him, keeping something like that from us?" ("Beside the Dying Fire"). Rick's reaction is surprisingly irascible: he replies "I didn't ask for this!" and even remarks that concerning Shane's death, "my hands are clean" and thus, denies any responsibility. It appears as if Rick wants to manipulate the whole group into bearing the burden of what happened, to turn the former promise of "we are all in this together" into another threat. Rick plays with the group's fear of being left alone and uses it to his advantage. Spitefully, he adds, "Maybe you people are better off without me, go ahead," well aware that he cannot simply abandon the group, or rather, that the group will not be able to function without him. The added "You stay, this isn't a democracy anymore" finally leaves no doubt that Rick is willing to forcefully get his own way within the group. Shane, it seems, has brought forth a side of Rick that the sheriff seemed unaware of. The treason and the breach of trust committed by his best friend (and on a second level, by his wife) ensure that Rick bids farewell to "officer friendly" for a while. It is not surprising that Shane's ghost will literally continue to haunt him at various times.[4] Shane thus functions as an emotional trigger for Rick. Whereas in his professional life Rick only had to apply the bad cop-method to strangers and vigilantes, he is now forced to use it on someone close to him, something that from this point on he is far more often willing to do.

Michonne and the Stereotype of the Angry Black Woman

Michonne (Danai Gurira) is not a classic adversary, but she is perceived as such by Rick. In the beginning, she is presented as the direct opposite of him and especially as the opposite of the other female characters. While most of them are chatting damsels in distress, Michonne is a taciturn, always scowling and katana-wielding amazon, who turned her now undead boyfriend and his best friend into a shackled zombie defense system,[5] used to distract other walkers. This emphasizes how solution-centered her thinking is, much like Rick's. Just like him, she used to be an "arm of the law," as she worked as a lawyer before the collapse; yet, unlike Rick she never introduces herself as such, as her former job is part of a life that does not exist anymore. However, Rick does not treat her like an equal, but reduces her to an "angry black woman," a stereotype that, as J. Celeste Walley-Jean argues, "arises from this foundation of negative images and the position of subordination of African American women that seeks to restrain their expression of anger by negatively labeling it" (Walley-Jean 71). Indeed, whenever Michonne encounters a new group of people, be it Rick's or the people of Woodbury under the control of the Governor (David Morrissey), she seems to take longer than others to be integrated, mainly because the male leaders seem to distrust a woman that is able to survive so long on her own.

Michonne first appears in the finale of season two, where she rescues Andrea who had been separated from the group after the downfall of Hershel's farm. However, she only encounters Rick in season three, episode seven ("When the Dead Come Knocking"). Shot in the leg by Merle (Michael Rooker), she appears wounded at the gates of the prison where Rick and his group found shelter, while Andrea, although she later on provides Rick with formula for Judith, remains loyal to the Governor and thus in Woodbury. Instead of talking to her, Rick stares at her and Michonne has to defend herself against the approaching walkers until she collapses and Carl saves her at the last minute by shooting several attacking walkers. She is carried into the prison and once she regains consciousness, Rick immediately asks her who she is. It is Michonne's reflex to grab for her sword, but Rick presses her to the ground and kicks it away, telling her, "We're not gonna hurt you unless you try something stupid first" ("When the Dead Come Knocking"). Rick's interrogation method, which consists of trying to gain her trust while simultaneously showing her that he is in charge of the situation, fails. Michonne refuses to tell him her name and only replies, "I didn't ask for your help."[6] Shortly after, Rick continues to interrogate her. Although he promises supplies and medical care, he does not hesitate to pinch it when she does not answer his questions quickly enough. While Rick's former good cop-attitude, i.e., caring for people in need, still shines through, he immediately switches into bad cop-mode to make Michonne talk. Yet this woman is unlike other people; she does not let herself be intimidated and threatened. Therefore, Rick tries again with a less demanding "You came here for a reason," and finally Michonne tells him about Woodbury and the Governor ("When the Dead Come Knocking").

After Shane's betrayal, Rick appears unable to empathize, especially with strangers that do not react as he expects them to. Additionally, Rick seems to be unable to make sense of Michonne because she is so much unlike the women surrounding him. Where Andrea, Lori, and even Carol in the beginning react mainly based on their emotions, Michonne remains a rational (and to him unfeminine) thinker. She, therefore, must be interrogated like a criminal to be made sense of. Furthermore, unlike most male characters, Michonne never actively challenges Rick's leadership, although she would be perfectly capable of leading the group. Yet, she challenges Rick's ego because her survival does not depend on him; she is able to survive on her own and even saves others while doing so. Surprisingly, Michonne later gains Rick's trust, even if she plays a key role in the conflict with the Governor. She can see through his crumbling facade after Lori's death and does so without applying the good cop/bad cop-technique. This does not mean that Rick's approach of identifying a person's true character is completely useless, but it proves that it is not without fault. Once Rick is in bad cop-mode, he makes wrong decisions, such as exiling Carol. Michonne cannot stop him from doing so, but she understands that bad cop-Rick is only a role he assumes as a mechanism of self-protection. In a sense, the development she undergoes is the opposite of Shane's; she starts as Rick's (presumed) antagonist and becomes one of his confidants, and ultimately, his lover. She successfully defies the stereotype he imposed on her.

The Governor, Cut from a Similar Cloth

The Governor is introduced in season three, episode three and constitutes the first proper villain Rick and his group are confronted with. While the threats before came either from the outside in form of nameless zombie hordes, or from inside of the group, in the form of Shane, the image of the enemy is now bundled into a man who prides himself on never giving his real name (only to do so one episode later).[7] Likewise, he is "a callback to the old authority, and a hope for the future" (Wilson). His name, like Rick's former occupation, suggests that he is in charge and, at first sight Woodbury, a small town with orderly model houses and high, guarded walls, suggests that he is, like Rick, good at his job. Yet appearances are deceiving. Although he seems composed on the outside, the Governor has a dark side which reveals his disturbed state of mind, or maybe even his true nature.[8] For the unknowing inhabitants of the town the Governor is a benefactor who provides them shelter. He governs Woodbury by constantly reminding them that they cannot survive outside its wall, i.e., outside his territory. Nobody is openly threatened, but the truth is always modified to make the inhabitants admire him, which causes Michonne to diagnose him with a "Messiah complex" ("I Ain't a Judas"). His twisted mind, however, becomes obvious in a variety of situations where he quickly switches between playing good and bad cop. At one point Merle abducts Glenn and Hershel's daughter Maggie (Lauren Cohan) to Woodbury. The Governor decides that Merle shall torture Glenn to get them to tell where the prison is, while he interrogates her. When he enters the room he asks a terrified Maggie, "May I?" ("When the Dead Come Knocking") when taking the chair opposite her and replies with "Thank you" when she agrees. He therefore commences the interrogation with a typical good cop-move to gain her trust, only to threaten rape moments later when she does not comply. Like Rick, he switches back and forth between the roles he assumes to unnerve his opponents.

Before he meets Rick, the Governor first considers Michonne to be his adversary since she stands out by being consistently suspicious of him and Woodbury in general. Her distrust is based on being disarmed which in the eyes of the Governor is necessary because Michonne's katana constitutes a penetrating and thus phallic weapon in the hands of a woman. When it is locked away in one of his cabinets, the patriarchal order is temporarily restored, and the Governor's power is, at least for the moment, not threatened. This instant is reminiscent of Rick's first encounter with Michonne. Interestingly, both men claim that they only take away her weapon because she is in safety already; however, in both cases safety means under their control. The good cop technique, i.e., trying to persuade her to give up her defense by reasoning her into believing that there is no danger, fails repeatedly. In this context it needs to be pointed out that the concept of the zombie, with its origins in Haitian folklore, is also strongly connected with images of colonialism and slavery (cf. Cortiel 189). With this in mind, Rick and the Governor appear as post-apocalyptic doppelgangers of colonizers who ruthlessly battle for space and male dominance. When Michonne kills Penny (Kylie Szymanski), the Governor's zombiefied daughter he kept hidden in a cage, and stabs his eye, she castrates his gaze[9] and makes it clear that initially she is more dangerous to him than Rick because she not only questions his authority, but further takes away his role as father and never grins and bears it. Michonne can first escape his wrath, but as soon as he learns that she is at the prison, and therefore in Rick's territory, his vindictiveness spreads from the individual Michonne to Rick's whole group. Andrea, as a link between both camps, arranges a meeting between both men, which functions as a mirror scene. The Governor demands that Rick surrenders Michonne in return for peace. Rick challenges him by uttering, "I thought you take responsibility" ("Arrow on the Doorpost"). The Governor replies, "I thought you were a cop, not a lawyer," and Rick finishes the argument with "Either way I don't pretend to be a governor." The irony is strong here because at the end of the day Rick does exactly this: he governs his group. Although the viewer is never informed about the Governor's concrete occupation before the collapse, it is striking how well he can keep up with Rick's cop-rhetoric. As Rick manages again and again to motivate his group, the Governor repeatedly manipulates people into trusting him, with fatal consequences for both sides. After a short redeeming arc, which (questionably) portrays the Governor as a caring ersatz-father, he declares war on Rick, this time over the ownership of the prison and he takes Michonne and Hershel as hostages. In a remarkable scene, Hershel addresses him with "Governor" ("Too Far Gone"), to which he replies, "Don't call me that."

Like Rick later on, he denies his own title, as if it frees him of all responsibilities. In the end, the Governor may be Rick's greatest threat because he is just like him. When the Governor's finally defeated, it is consequently not by Rick, but through the hands of Michonne and Lilly (Audrey Marie Anderson). This again can be interpreted as another hint at Rick and the Governor's similarity. They are so similar that they cannot kill each other because it would be like killing themselves. In the aftermath of the war the group is split up again and Rick escapes, barely alive, with Carl. It is then his own son who accuses his father, "You were the leader" and "Now you're nothing" ("After"). The good cop turned bad cop is now no longer a cop at all. In the face of an antagonist who is too similar to him his former strength, to make people either trust or obey him, loses all power.

Deanna Monroe and the Failure of Good Nature

Deanna (Tovah Feldshuh) first appears in season five, episode twelve when Rick and his group are brought to Alexandria, a democratic, gated community seemingly untouched from the world outside, with nice houses and friendly, clean people who in part have never left its walls. A former congresswoman, Deanna mastered the rhetoric of appealing to people and being accepted as their leader without intimidating them, similar to how Rick was in the very beginning. When the group arrives, Deanna asks them to conduct recorded interviews with them. Her interview with Rick provides interesting insight into his current mindset, as she asks him what he was before the collapse and he simply answers, "I don't think it matters anymore" ("Remember"). But Rick cannot deny his past; he tells Deanna the truth, and she immediately employs him as Alexandria's new constable, with Michonne to assist him. Her main function in the series is thus to get Rick back on his cop track. Once back in a uniform, Rick's ego recovers dangerously quickly, which becomes obvious when he attacks the husband of Jessie (Alexandra Breckenridge), his new love interest.[10] Still, Deanna trusts him, although Rick repeatedly, and aggressively, questions her mild leadership style. When Alexandria threatens to crumble under the pressure of a massive walker horde outside and the town is simultaneously attacked by a gang of murderous, feral people, Rick feels vindicated in his assumption that a friendly leadership cannot provide safety. Deanna is no governor; she is a planner who wants to rebuild humanity in peace, an intention the Rick of the early seasons would have appreciated; but, the current Rick, with all his calamities suffered, can no longer believe in. Deanna is thus unable to turn from good cop to bad cop and therefore, is neither able to save Alexandria, nor survive. Her failure is, shortly before her death, denounced by her son Spencer (Austin Nichols), who accuses her, "You're the reason we're so screwed, you made us this way!" ("Now"). While Rick's turn from good cop to bad cop initially appeared excessive and unjust, it now becomes obvious to be a necessary means. His former, over the course of the series constantly repeated credo "We don't kill the living" has become redundant, as the living cause more trouble than the dead, as the inhabitants of Alexandria, and especially Deanna, have to experience. Indirectly, she makes him a leader again by showing him that going back to being good only is not an option as it results in failure. The group needs someone who guides them, even if this means to make morally questionable decisions. Rick's personal struggle between good and bad, between saving and killing people, is thus what made them survive for so long.

Conclusion

Rick's journey in *The Walking Dead* not only describes a man's search for a safe place to provide shelter for his family and friends, it also chronicles the decay of humanity in extreme circumstances. A former poster boy of good nature and humanity, Rick learns in the course of the show that he has to give up his moral standards in order to survive. At the same time the fragility of sanity becomes obvious. With every new challenge, Rick has to move further away from being everybody's friend and helper to an authority that is either accepted or feared. By doing so, the borders between a good and a bad cop are constantly blurred. His former job remains the only fixed constant in his life, but this is both burden and blessing. His title raises certain expectations in people, good and bad ones, expectations are dumped on him only to praise or denounce him, depending on the outcome of the situation.

Simultaneously, Rick forces such behavior by repeatedly emphasizing his role within his group, not only through his actions but also by dressing and talking like a cop. In the end, it is his irritating and at times unpredictable behavior which successfully saves the others. His use of the good cop/ bad cop-technique can therefore be seen as a necessary evil, a remnant from a civilization that as such no longer exists but still proves to be useful. The adversaries, and in some cases not all of them turn out to be proper ones, he encounters along the way accelerate his development in good and bad ways but the all have one thing in common: they ensure that the role of the cop remains important, even if the law itself no longer exists.

Notes

1. AMC used the tagline "Fight the Dead. Fear the Living" to promote season three of the series.
2. "Guts," episode 1.2, 7 November 2010. The person asking Rick who he is is Merle, Daryl Dixon's brother.
3. In episode three of the same season, "Vatos," Rick is confronted with a group of Latinos who first steal his bag and later kidnap Glenn, the pizza delivery-boy who saved his life. The encounter almost turns violent as Rick threatens to hurt one of the Latinos if they do not return Glenn and the bag, he switches from the good cop trying to protect "his" group to the bad cop threatening violence if he does not get what he wants.
4. Shane appears, armed and seemingly aiming at Rick, in one of his hallucinations during the battle at Woodbury (c.f. "Made to Suffer," episode 3.8, 2 December 2012).
5. The image of the two black men being pulled along with chains around their necks alludes to slavery and how black people were used as commodities and not actual people.
6. It is important to note here that although Michonne meets Rick after the Governor, her "receptions" in Woodbury and at the prison are eerily similar, both times the men take away her most precious possession, the katana.
7. Although he tells Andrea that his name is "Philip Blake," and she from then on continues to call him by his real first name, in the following he will still be referred to as The Governor because this is how he is recognized as by the majority of characters.
8. For example, he maintains a secret chamber with aquariums full of walker heads,

kills a group of military personnel and later on kills the remaining Woodbury survivors, among other things.

9. In this context it is worth mentioning that in the comics Michonne castrates the Governor in revenge for him raping her, she literally strips him off his manhood.

10. This mirrors the love triangle between Rick, Lori and Shane and therefore hints at the fact that Rick has not yet overcome his bad cop-only phase.

Works Cited

Berk, Isaac. "*The Walking Dead* as a Critique of American Democracy." *CineAction* 95 (2015): 48–55. Print.

Bishop, Kyle William. *American Zombie Gothic. The Rise and Fall (and Rise) of the Walking Dead in Popular Culture.* Jefferson, NC: McFarland, 2010. Print.

Conrad, Joseph. *Heart of Darkness & Other Stories.* Wordsworth Classics. Ware: Wordsworth Editions Limited, 1998 [1899]. Print.

Cortiel, Jeanne. "Travels with Carl: Apocalyptic Zombiescape, Masculinity and Seriality in Robert Kirkman's *The Walking Dead.*" *The "Journey of Life" in American Life and Literature.* Ed. Peter Freese. Heidelberg: Universitätsverlag Winter, 2015. 187–204. Print.

Gavaler, Chris. "Zombies vs. Superheroes: *The Walking Dead* Resurrection of Fantastic Four Gender Formulas." *ImageTexT: Interdisciplinary Comics Studies* 7.4 (2014). Print.

Rafaeli, Anat, and Robert I. Sutton. "Emotional Strategies as Means of Social Influence: Lessons from Criminal Interrogators and Bill Collectors." *The Academy of Management Journal* (1991): 749–775. Print.

Sugg, Katherine. "*The Walking Dead*: Late Liberalism and Masculine Subjection in Apocalypse Fiction." *Journal of American Studies* (1 November 2015): 793–811. Print.

The Walking Dead: Season One. Prod. David Alpert, Frank Darabont, Charles H. Eglee, Gale Anne Hurd and Robert Kirkman. AMC. 2010. DVD.

The Walking Dead: Season Two. Prod. David Alpert, Frank Darabont, Gale Ann Hurd, Robert Kirkman and Glen Mazzara. AMC. 2011. DVD.

The Walking Dead: Season Three. Prod. David Alpert, Gale Ann Hurd, Robert Kirkman and Glen Mazzara AMC. 2012. DVD.

The Walking Dead: Season Four. Prod. David Alpert, Scott. M. Gimple, Gale Ann Hurd, Robert Kirkman, Tom Luse and Greg Nicotero. AMC. 2013. DVD.

The Walking Dead: Season Five. Prod. David Alpert, Scott. M. Gimple, Gale Ann Hurd, Robert Kirkman, Tom Luse and Greg Nicotero. AMC. 2014. DVD.

The Walking Dead: Season Six. Prod. David Alpert, Scott. M. Gimple, Gale Ann Hurd, Robert Kirkman, Tom Luse and Greg Nicotero. AMC. 2015. Television.

Walley-Jean, J. Celeste. "Debunking the Myth of the 'Angry Black Woman': An Exploration of Anger in Young African American Women." *Black Women, Gender & Families* 3.2 (2009): 68–86. Print.

Wilson, Steven Lloyd. "'The Walking Dead': Anti-Libertarian Critique." Salonwww. Web. 15 Oct. 2015.

"The girl needs a little monster in her man"
Heroes and Villains in the Works of Joss Whedon

DON TRESCA

In Christopher Golden and Nancy Holder's *Sons of Entropy*, book three of *The Gatekeeper Trilogy*, a series of *Buffy the Vampire Slayer* novels, the character of Rupert Giles, Buffy's mentor and guide, states, "I find it most remarkable that we, who are so intimately involved in the battle between good and evil, are even more involved with the shades of gray in between them" (160). That quote perfectly sums up the relationship many of Joss Whedon's characters have with the moral dilemma of good versus evil. Each of Whedon's television shows, *Buffy the Vampire Slayer* (1997–2003), *Angel* (1999–2004), *Firefly* (2001–2002), *Dollhouse* (2009–2010), and *Agents of S.H.I.E.L.D.* (2013–present), and his films, both those he has directed—*Serenity* (2005), *Dr. Horrible's Sing-Along Blog* (2008), *The Avengers* (2012), *Much Ado About Nothing* (2013), and *The Avengers: Age of Ultron* (2015)—and those he has written (or co-written)—*Toy Story* (1995), *Alien: Resurrection* (1997), *The Cabin in the Woods* (2012), and *In Your Eyes* (2014)—have, as a central theme, both the conflict between good and evil and the interconnectivity between the two moral extremes. Characters throughout Whedon's work drift in the gray area between good and evil, becoming heroes one day and villains the next. These ambiguous hero/villains challenge the audience's view of traditional morality and how this view reflects upon our modern society. Many of Whedon's characters shift between good to evil or vice versa with such regularity that it demonstrates how the definitions of these once-polar opposites are no longer as clear cut as they were once perceived to be and how sometimes, in order to emerge victorious, the man and the monster must become one. While I will be referencing numerous characters throughout this essay that cross this once uncrossable line, the primary characters I will focus on are Angel and Spike (from *Buffy the Vampire Slayer* and *Angel*); Boyd Langdon (from *Dollhouse*), and Agent Grant Ward (from *Agents of S.H.I.E.L.D.*).

Ironically, given Whedon's proclivity to create ambiguous hero/villain amalgam characters, many of the genres he chooses to work in are amongst the ones traditionally most stark in their division of good and evil. Vampire, western, space opera, and superhero narratives typically depend on a strong contrast between the heroes and villains of the text. A clearly defined evil (a vampire lord, a ruthless band of black-hatted outlaws, an intergalactic empire bent on subjugation and enslavement of defenseless planetary systems, or a megalomaniacal super-villain) threatens a group of innocents. In response, a force of good arises to intervene against the evil. This good protector is always fundamentally better than the mere victims with self-selected organizations, methods, and values (Knehans 7). Eventually after various trials and battles, good is triumphant, and the victims are saved from the savage predation of the now-vanquished evil.

However, in Whedon's work, he seems to recognize a fundamental characteristic of villains that most traditional stories tend to overlook. Nobody ever sets out at the outset of his or her life seeking to be the villain of the story. The villain begins its journey in the same way the hero does, wanting something and being unable to achieve it for whatever reason. For Whedon, the most dangerous of all the villains are those who are ultimately unaware of their own dark natures: "Nobody thinks that they're a bad guy. Nobody thinks that they're not righteous. I've dealt with people that are truly villainous ... people who have done appalling things to other people on purpose. And they think they're righteous" (Riess 116). This quote seems to encapsulate Whedon's approach to the hero/villain dichotomy in his work. The lines between the two extremes blur because villains engage in righteous actions for despicable reasons while heroes engage in despicable actions for righteous reasons. What typically sets the villain apart from the hero is the methods he or she employs to achieve his or her desire (Johnson 41). Villains use force, threats, deceit, and manipulation to succeed in their goals. Yet, in Whedon's work, many of his "heroic" characters find themselves using these same methods in the cause of good. It is telling that Whedon's first foray into direct adaptation of Shakespeare for the screen is *Much Ado About Nothing* since Shakespeare himself used that play to reveal the ways in which the morally-dubious activity of deception could be used to achieve a positive outcome, namely the marriage of Benedick and Beatrice (Henze 188), while at the same time showing how deception can also be used negatively as when Claudio uses deception to ruin the life and reputation of Hero (Henze 200). Likewise, we see numerous examples of this in Whedon's other works. For instance, Angel and Buffy trick Faith into revealing the Mayor's master plan by deceiving her into believing that Angelus, Angel's evil half, has returned in the *Buffy* episode "Enemies." Another example is when the crew of *Serenity* disguises their ship as a Reaver ship to get through the Reaver blockade in *Serenity*, and finally, Nick Fury deceives the Avengers about Agent Coulson's death to bring them together as a team in *The Avengers*.

Thus, Joss Whedon's universe is one of grays rather than crisp black and white oppositions (Abbott 1). For Whedon, evil is not an external reality that can be easily dispatched or conquered. It is an internal part of each person living on the planet. Even the character that represents the purest of Whedon's heroes, his Vampire Slayer Buffy, was created from evil when the first Watchers used the essence of a demon to infuse the body of a young girl with the power to destroy the forces of darkness in the world (as seen in the *Buffy* episode "Get It Done"). We as humans are both spiritual and carnal, altruistic and selfish, magnanimous and narrow-minded, in short both good and evil (Riess 115). Our darker self, what Carl Jung called our "shadow self" (Jung 77), is within us always and to deny this is destructive to both ourselves and everyone around us. Whedon shows us this time and again with Angel/Angelus' destructive reign of terror during *Buffy* Season Two and *Angel* Season Four, Dark Willow's near-destruction of the world at the end of *Buffy* Season Six, and the hubris of Tony Stark and Bruce Banner in their creation of Ultron, the megalomaniacal robotic villain in *The Avengers: Age of Ultron*.

Two of the first major characters in one of Whedon's works to fully explore the sophisticated dynamics of his moral universe were Angel and Spike, the vampires with human souls. Angel was introduced in the first episode of *Buffy*, "Welcome to the Hellmouth," but his vampirism wasn't revealed until the seventh episode of the first season, appropriately titled "Angel." Despite the fact that, up until the point when his vampirism is revealed, Angel is presented as an ally, albeit a somewhat reluctant one at times, Angel's vampirism creates a dilemma for Buffy. Her sacred moral duty is to slay vampires to protect innocent humans, yet Angel seems to have nothing but good and noble purposes and has not once posed a threat to an innocent person. Therefore, is there room in her black-and-white moral duty to slay vampires for the possibility that some vampires may not need to be slain? She presents the situation ("Can a vampire ever be a good person?") to her mentor/Watcher Giles who responds, "A vampire isn't a person at all. It may have the movements, the memories, even the personality of the person that it took over, but it is still a demon at the core. There is no half-way" ("Angel"). Angel, however, it is discovered, is a unique situation in that he had his human soul restored to him via a gypsy curse; therefore, he is no longer, as Giles put it, "a demon at the core." However, Angel is never presented in either *Buffy*, or his own show *Angel*, as morally perfect. Both shows take great pains to indicate that having a human soul does not make one inherently good and noble. Characters like Warren (in season six of *Buffy*) and Lindsey McDonald (throughout *Angel*'s five season run) are prime examples of ensouled human characters with deeply flawed moral values. At the same time, the show presents supposedly "soulless" demons that are capable of great morality, such as Whistler and Clem from *Buffy* and Doyle and Lorne from *Angel*. Angel himself, for his part, is revealed to have done some horrible things even during the times after he was granted his soul, including leaving a hotel full of innocents in the hands of a fear demon (in "Are You Now, or Have You Ever Been?"); trapping a group of lawyers (albeit evil lawyers) in a cellar with two vicious vampires (in "Reunion"), transforming an innocent sailor into a vampire (in "Why We Fight"), and being tempted on at least two separate occasions (in "Angel" and "Orpheus") to drink the blood of an innocent. Buffy's reluctance to slay Angel (just as her later reluctance to slay Spike once he has become "neutered" with the Initiative chip and then gains his own soul at the end of *Buffy* season six) has less to do with the black-and-white humans vs. vampires moral certainty (that is, all vampires and demons are a threat to humanity and, therefore, must be slain before they can act on that threat) and a more sophisticated morality in which each action must be carefully weighed against the consequence that action may invoke (Greene and Yuen 12). Buffy's experiences with Angel and Spike and her exposure to this moral ambiguity leads Buffy to occasionally even hesitate to kill a potentially dangerous foe outright (such as her episode-long battle with Holden in "Conversations with Dead People") to evaluate whether or not the creature she is fighting deserves to die.

In the *Buffy* season four episode "The Initiative," the viewers learn that Spike, a recurring master vampire villain whose first appearance was in the season two episode "School Hard," has been captured by the Initiative, a military organization headquartered below the UC Sunnydale campus. The Initiative's goals are to capture and study unknown "hostiles" (vampires and demons) and render them incapable of harming humans. Spike is outfitted with a special behavior chip that makes him incapable of hurting or killing anyone without causing him excruciating pain. Spike has tried to kill her and her friends numerous times, yet Buffy refuses to slay Spike once she learns that he has been "neutered," despite the fact that Spike is still clearly evil and, at one point, allies himself with the season four Big Bad, a human-demon hybrid named Adam. Several Whedon scholars have hypothesized on exactly why Buffy seems incapable of slaying Spike despite the fact that he is an evil vampire, even when rendered harmless by the Initiative chip. Rhonda V. Wilcox sees in Spike a philosophically different approach to the nature of good vs. evil than she sees in Angel. While Angel does good because he possesses a human soul and can therefore be seen to represent an essentialist definition of good in that he does good works because it is an essential part of his humanity, Spike must choose to be good despite his evil nature, showing that even the worst among us can be capable of moral change given the right circumstances. Therefore, Spike can be seen to represent an existentialist definition of good in that he must overcome his dark nature and choose good, not because it is an essential part of his nature but because it is the correct thing to do (Wilcox, "Every Night I Save You" 88).

Spike's primary motivation to do good is his realization of the romantic feelings he has for Buffy. He strives to do good to prove himself worthy to her. Despite his "evil" and selfish nature, he constantly chooses the interests of Buffy over his own. He protects the secret of Dawn's (Buffy's sister) identity as the Key from the hell goddess Glory even while undergoing torture and near-death. He does this not for reward or redemption but simply because he is allowing "love [to] command his behavior" (Bradney 4), in much the same way he reacts by weeping bitterly at the sight of Buffy's dead body in "The Gift." He continues to fight the good fight even after Buffy's death, aiding the rest of the Scooby gang in maintaining the ruse that Buffy still lives in order to keep the vampire population of Sunnydale cowed, to prove his unending love for her (Crusie 90). Spike becomes the living embodiment of one of Whedon's most enduring themes, the power of love to overcome even the darkest of evils (Tresca 133). Spike, motivated by his love for Buffy, has actually cultivated his own soul, built an inherent goodness within himself that overshadows his natural evil impulses (Milavec and Kaye 179).

However, Spike learns that even love and the best of intentions cannot completely eradicate a creature's innate evil. Marti Noxon, the executive producer of *Buffy the Vampire Slayer*, stated that Whedon and his writing staff knew that Spike was eliciting audience sympathy because of his love for Buffy and that many viewers were losing touch with the fact that Spike was, at his core essence, still an evil soulless vampire (Gottlieb 3). So Whedon and his staff countered the audience sympathy in a dark and violent manner, by having Spike attempt to rape Buffy in the episode "Seeing Red." The rape scene became the focal point of much discussion amongst Buffy scholars with many agreeing with Whedon's decision to use the near-rape as a primary motivating factor for Spike to complete his redemption arc by battling for and ultimately achieving his soul (Symonds, "A Little More Soul Than Is Written" 14) while others saw the scene as a betrayal of the show's portrayal of Spike's redemption up to that point with some even suggesting that the "rape" was justified by suggesting that Buffy was sexually abusing Spike throughout the whole of season six and that her refusal to acquiesce to his sexual advances in the rape scene were part of the abuse cycle (Smirnov 10). Gwyn Symonds even points out that the scene is shot in such a manner (with the camera alternating between close-ups of Buffy and Spike separately) as to "encourage a complex audience engagement with both ... the perpetrator and the victim," to reinforce the audience's shifting empathy with both Buffy and Spike ("Solving Problems with Sharp Objects" 148).

Ultimately, however, Spike's story becomes a heroic one, a tale of a man trying to be better than he is supposed to be. The story is attractive because it details the inspirational possibility inherent in the fact that we do not have to be defined by how we start out or how others see us (Symonds, "Solving Problems with Sharp Objects" 146). After his near-rape of Buffy makes Spike realize that he is still as internally corrupt as he always has been, he makes a pilgrimage across the world to quest for his soul. He endures great pains and trials, many which nearly kill him, to achieve his goal of giving Buffy "what she deserves" (in "Two to Go" and "Grave"). Unlike Angel, who had his soul forced upon him with a gypsy curse, Spike fights for his soul and ultimately succeeds in his quest. Spike, as a soulless vampire, made himself unique in the Whedonverse by asserting his existential prerogative to seek a fundamental change in his being (McLaren 20). While the Initiative chip that was forced upon him (in much the same way Angel's soul was forced upon him) was designed to prevent evil action, it did not prevent Spike from approaching many of his choices (particularly his alliance with Adam in *Buffy* season four) from an evil morality, even his ultimate decision to ally with Buffy and her friends. However, with his soul, although he ultimately allies with Buffy again, he does so from an entirely different moral perspective, from that of the morally good. Spike deliberately chooses to have a soul knowing that the soul will allow him to make moral determinations solely of his own free will (McLaren 22).

After *Buffy* and *Angel* ended their series run (and *Firefly* failed during its initial first season), Joss Whedon met with actress Eliza Dushku (who portrayed the dark slayer Faith in several seasons of *Buffy* and *Angel*) to develop ideas for a new television series to pitch to the Fox television network. That series, *Dollhouse*, became something totally different from any of Whedon's previous projects. A modern-day science-fiction show, *Dollhouse* told the story of a secret organization called the Rossum Corporation which used technology to remove a person's mind and personality from his or her body and replace them with the mind and personality of another, thus allowing certain individuals (known collectively as the Dolls) to be used as virtual sex slaves, allowing them to be imprinted with selected identities that matched the specifications of the Corporation's clientele.

One of the dolls, codenamed Echo (played by Dushku), has a "flaw" in her programming, allowing her to block an imprint and retain some independent thought, which, as the series progresses, allows her to begin to corrupt and dismantle the Corporation's immoral business from the inside. One of her earliest allies is her handler, Boyd Langton, a former police detective and the so-called "moral center" of the show, according to Joss Whedon himself in his DVD commentary on the episode "Man on the Street." Boyd cares for Echo, even risking his life for her by taking a bullet meant for her in "The Target." He also appears to question the Dollhouse at every opportunity, questioning their methods and their moral decisions of sending the Dolls out on certain assignments. However, in a twist, Boyd is revealed in the second season episode "The Hollow Men" to be the head of the Rossum Corporation and that he was merely manipulating Echo to use her unique "flaw" to design a method of making corporation executives immune to imprinting while enslaving the rest of the population through the imprinting technology. Fortunately, Echo and her allies are able to overpower Boyd before he can harvest her spinal fluid to make his immunity serum. Echo wipes Boyd's memory with the imprinting technology, turning him into a Doll. She then uses him as a suicide bomber to destroy Rossum's imprinting technology.

Although Boyd's deception and betrayal of Echo and her allies is shocking, Daniel A. Clark and P. Andrew Miller suggest that the audience should have been aware of Boyd's duplicity from the beginning. Boyd is presented as an authority figure, an ex-cop and Echo's "father figure." Authority figures in Whedon's work are almost always either corrupt from their first appearance, like Mayor Wilkins in the third season of *Buffy* or General Talbot in *Agents of S.H.I.E.L.D.*, or are eventually revealed to be corrupt, such as Gwendolyn Post in the *Buffy* episode "Revelations" or John Garrett in the *Agents of S.H.I.E.L.D.* episode "Nothing Personal." This corruption of authority and power in the Whedonverse ultimately means not only a loss of morality and ethics but the loss of one's soul (which happens literally to Boyd when his mind is wiped with the imprinting technology) (Clark and Miller 16).[1] Boyd's betrayal literally acts as a re-imprinting of the show's morality. The man that the viewers thought Boyd was did not die; he never truly existed in the first place (Wilcox, "Echoes of Complicity" 14). Returning to Whedon's comment about Boyd serving as the show's "moral center," the realization of that quote becomes clear. Boyd is the "moral center," but the moral center is absolutely corrupt. Although there are morally upstanding characters within the show (such as FBI Agent Paul Ballard and Caroline, Echo's pre-imprint identity), even they are eventually corrupted by the inherent immorality surrounding the Dollhouse. Ballard becomes part of the system when he agrees to become Echo's new handler when Boyd is "promoted" to head of security, and Caroline becomes so corrupted by her own morally ambiguous behavior (as an animal rights' activist whose actions lead to the death of one of her allies) that Echo refuses to return to that personality and chooses to keep her imprinted one at the series' conclusion.

Boyd's eventual betrayal of Echo and her allies calls into question everything seen in the show prior to this point. Whedon and his writers did everything in their power to ensure that the audience developed a trust in Boyd almost as strong and unwavering as Echo's, to make the revelation of his duplicity that much more devastating to the viewers. Hearn's sexual abuse of his charge Sierra in "Man on the Street" revealed just how much trust a Doll was programmed to have for his or her handler, as well as how easy it was for that trust to be exploited by an immoral handler like Hearn. Simultaneously, it affirmed the audience's trust and belief in Boyd by placing him in sharp contrast with Hearn throughout the episode (Souza 206). Boyd was presented throughout the series as a fatherly type, strong, protective, and caring. He had a trustworthy face, a knowing smile, and kind eyes, features society is taught to believe connote a man of goodness and nobility. By presenting the character of Boyd in this way, Whedon reveals how close the Dollhouse is to the real world outside the show, how everyone is "programmed" to project certain traits and ideals onto someone based on his or her appearance alone. Given a fictional world in which it was very hard to trust or relate to any of the characters, Boyd was seen as the everyman, the one to latch onto. By then revealing Boyd to be the complete opposite of what the viewers thought that he was, Whedon revealed how similar they were to clients and staff of the dollhouse (Souza 211–212).

The final character I wish to examine in this discussion of the good/evil spectrum in Whedon's work is Agent Grant Ward from *Agents of S.H.I.E.L.D.* Ward is really the first character viewers meet in the pilot episode of the series as he is promoted to Level 7 clearance in order to meet with the recently-resurrected Agent Phil Coulson who has been tasked to head a new team of S.H.I.E.L.D. agents to investigate various crises around the world related to superhuman activities. Ward initially appears to be the traditional stereotypical action hero: strong, handsome, brave, resourceful, good with both a variety of weapons and hand-to-hand fighting techniques. Ward even performs several death-defying heroic deeds to save key members of the team, diving from a moving airplane with an unsecured parachute to save Simmons in "FZZT" and protecting Fitz from hostile gunfire after learning that the two of them were sent on a literal "suicide mission" in "The Hub." When it is revealed that Ward is actually a Hydra agent planted within S.H.I.E.L.D. as part of the villainous organization's plan to bring about world peace through assassination and intimidation (as revealed in the film *Captain America: The Winter Soldier* [2014]), Whedon showed, as he did with Boyd in *Dollhouse*, how a noble and heroic façade can mask a dark nature. Like Boyd, Ward was a villain from the beginning, using his skills and heroic persona to fool both the characters in the show and the audience watching the show into believing he is something that he is not. Again, Ward shows how easily an audience can be "programmed" into accepting a villainous character as a hero by using the tropes of the traditional action hero to build Ward's character.

Because Ward's actions put him into opposition to Agent Coulson and his team of S.H.I.E.L.D. agents, he is marked immediately as a antagonist, as a villain, yet Fernando Gabriel Pagnoni Berns and Cesar Alfonso Marino showed how closely aligned the goals of many of the villains of Whedon's works are with the goals of their heroic adversaries. Hydra's goal (as revealed in *Captain America: The Winter Soldier*) is to create a world of peace, although it is shown that this can only be done by using the hijacked S.H.I.E.L.D. helicarriers to assassinate millions of innocent people. The goal of universal world peace and order is one shared by numerous Whedon villains, including Jasmine in *Angel*, the Operative in *Serenity*, Doctor Horrible in *Doctor Horrible's Sing-Along Blog*, Loki in *The Avengers*, and Boyd in *Dollhouse*. Even the technicians and scientists in the Whedon-scripted *The Cabin in the Woods* do what they do (murder innocent young people in blood sacrifices to ancient gods) to "save the world." Because of the similar nature of many of their goals, good and evil face the philosophy of relativism, which turns them into "fragile preferences of a particular social group at a certain place and time" (Pollefeyt 129) rather than true opposing poles of morality (Berns and Marino 147). Ward is in many ways the perfect representation of this. His commitment to Hydra's ordinarily altruistic cause of world peace is so thorough that Eric Koenig's lie-detector test in "The Only Light in the Darkness" fails to detect Ward's villainous nature. Although his status as a villain is maintained permanently after the events of "Turn, Turn, Turn," Ward sees himself as a hero and occasionally plays the role, such as assisting Coulson's team in "The Frenemy of My Enemy" to infiltrate a Hydra base to rescue Skye. Unlike Spike, however, who transforms from a villain to a hero because of his love of Buffy, Ward is unable to use his love of Skye to transition from villain to hero although he claims that he wants to, indicating that his love allowed him to feel that "for the first time in a while I wanted something for myself" (as stated in "The Beginning of the End"), which indicates that he is starting to break his attachment to John Garrett and Hydra, but the tragedy is that Skye does not recognize this (Selene 5) and rejects his love. Unlike Spike, who has the strength of will to fight for Buffy's love rather than take her rejection to heart, Ward uses Skye's rejection to plummet deeper into evil, eventually forming a semi-romantic attachment to Agent 33, a disfigured former S.H.I.E.L.D. agent that is brainwashed into doing Hydra's bidding. The loss of Agent 33 at his own hands in "S.O.S." (although Ward blames Coulson and his team for her death) drives him deeper into despair and completes his transition into full villain mode which culminates with his own death and his return possessed by the spirit of the powerful Inhuman known as Hive (in "Maveth"). Ward becomes Whedon's representation of the dark nature of love which, when lost, turns the lover into a remorseless, bloodthirsty, vindictive killer (Tresca 141). Although many Whedon fans see Ward mirroring the Angel/Angelus storyline in *Buffy* because they see the S.H.I.E.L.D. Agent Ward persona as a cover that conceals the character's true villainous nature in the same way that Angel is merely a cover that conceals the true villainous nature of Angelus (Thomas 3),[2] Ward actually serves as the shadow opposite of Spike. Spike is a villain who becomes a hero because of love while Ward is a hero who becomes a villain also because of love (Thomas 8). But both characters represent a nebulous and flexible connection between the moral stances of good and evil, showing how a character may easily cross between the two.

The list of characters in Whedon's work that cross the boundaries between good and evil is practically endless. Fans and scholars point to numerous characters who engage in "hidden loyalties and betrayed trust" which code them as "evil," yet they are eventually welcomed back into the fold of the "good" chosen family with open arms (i.e., Angel, Jenny Calendar, Faith, Spike, Willow, Anya, Wesley, Mal Reynolds, Jayne, Tony Stark). It is rare for a character to be permanently disbarred from the "good" family unit once their betrayals become known, but it does happen (such as the examples of Boyd and Ward as discussed earlier in this essay) (Alvi 39). As we entered the new millennium, our heroes had to contend with a postmodern world fraught with ambiguities in which traditional systems of right and wrong, darkness and light, good and evil became chaotic and open to limitless possibilities. Joss Whedon is one of the few television and film auteurs right now willing to examine this new world order and to discuss this moral fluidity in his work in a way, that shows the basic interdependence between good and evil in a way that is dramatic yet profound. As Gregory Sakal describes it:

> Without this struggle [between good and evil], there would be no choices to make, no exercise of free will, no path to redemption.... [W]hat distinguishes humanity from many of the other creatures we meet in [Whedon's] world is the ability to choose.... Without the darkness, the light would go unrecognized. Without evil, there would be no struggle, no sacrifice, and hence no possibility of or need for redemption [251].

Good and evil, in the end, are behaviors, not ineffable Kantian categories. Like all behaviors, they are mutable and socially constructed. One person's hero is another's villain. What rules the world is neither good nor evil, but rather the balance between the two, the paradox that neither has meaning without the other. This paradox is at the crux of all of Whedon's work, whether television shows or movies, whether in a world of vampires and demons, of superheroes and spies, of spacemen and cowboys (Ramachandran 76).

Notes

1. Certain authority figures (i.e., Giles in *Buffy*, Captain Mal Reynolds in *Firefly/Serenity*, and Phil Coulson in *Agents of S.H.I.E.L.D.*) escape the corruption by creating an egalitarian structure to their hierarchical organization, treating those they have authority over as equals rather than inferiors.

2. This, of course, assumes that Angelus is the "true" character while Angel is merely the "cover" concealing that character, and that is not a view that all Whedon scholars share.

Works Cited

Abbott, Stacey. "Walking the Fine Line Between Angel and Angelus." *Slayage: The Journal of the Whedon Studies Association* 3.1(9), August 2003. Web. 20 Nov. 2014.

Alvi, Zalina. "Goliath Is People! How *Dollhouse* Took Distrust to a Whole New Level." *Inside Joss' Dollhouse: From Alpha to Rossum.* Eds. Jane Espenson and Leah Wilson. Dallas: BenBella, 2010. 35–45. Print.

"Angel." *Buffy the Vampire Slayer: The Chosen Collection.* Dir. Scott Brazil. Wr. David Greenwalt. Twentieth-Century-Fox Home Entertainment, 2005. DVD.

"Are You Now, or Have You Every Been?" *Angel: The Complete Second Season.* Dir. David Semel. Writ. Tim Minear. Twentieth-Century-Fox Home Entertainment, 2003. DVD.

The Avengers. Dir. Joss Whedon. Writ. Joss Whedon and Zak Penn. Walt Disney Studios Home Entertainment, 2012. DVD.

Avengers: Age of Ultron. Dir. Joss Whedon. Writ. Joss Whedon. Walt Disney Studios Home Entertainment, 2015. DVD.

"Beginning of the End." *Agents of SHIELD: The Complete First Season.* Dir. David Straiton. Writ. Maurissa Tancharoen and Jed Whedon. Walt Disney Studios Home Entertainment, 2014. DVD.

Berns, Fernando Gabriel Pagnoni, and Cesar Alfonso Marino. "Embracing Goodness (and Colorful Costumes) Amid a World of Gray." *The Comics of Joss Whedon: Critical Essays.* Ed. Valerie Estelle Frankel. Jefferson, NC: McFarland, 2015. 147–153. Print.

Bradney, Anthony. "'I Made a Promise to a Lady': Law and Love in *Buffy the Vampire Slayer*." *Slayage: The Journal of the Whedon Studies Association* 3.2(10), November 2003. Web. 20 Nov. 2014.

The Cabin in the Woods. Dir. Drew Goddard. Writ. Joss Whedon and Drew Goddard. Lionsgate Home Video, 2012. DVD.

Captain America: The Winter Soldier. Dir. Anthony Russo and Joe Russo. Writ. Christopher Markus and Stephen McFeely. Walt Disney Studios Home Entertainment, 2014. DVD.

Clark, Daniel A., and P. Andrew Miller. "Buffy, the Scooby Gang, and Monstrous Authority: *Buffy the Vampire Slayer* and the Subversion of Authority." *Slayage: The Journal of the Whedon Studies Association* 1.3(3), June 2001. Web. 20 Nov. 2014.

"Conversations with Dead People." *Buffy the Vampire Slayer: The Chosen Collection.* Dir. Nick Marck. Writ. Jane Espenson and Drew Goddard. Twentieth-Century-Fox Home Entertainment, 2005. DVD.

Crusie, Jennifer. "Dating Death." *Seven Seasons of Buffy: Science Fiction and Fantasy Authors Discuss Their Favorite Television Show.* Ed. Glenn Yeffeth. Dallas: BenBella, 2003. 85–96. Print.

Doctor Horrible's Sing-Along Blog. Dir. Joss Whedon. Wr. Maurissa Tancharoen, Jed Whedon, Joss Whedon, and Zack Whedon. New Video Group, 2008. DVD.

"Enemies." *Buffy the Vampire Slayer: The Chosen Collection.* Dir. David Grossman. Writ. Douglas Petrie. Twentieth Century-Fox Home Entertainment, 2005. DVD.

"The Frenemy of My Enemy." *Agents of SHIELD: The Complete Second Season.* Dir. Karen Gaviola. Writ. Monica Owusu-Breen and Paul Zbyszewski. Walt Disney Studios Home Entertainment, 2015. DVD.

"FZZT." *Agents of SHIELD: The Complete First Season.* Dir. Vincent Misiano. Writ. Paul Zbyszewski. Walt Disney Studios Home Entertainment, 2014. DVD.

"Get It Done." *Buffy the Vampire Slayer: The Chosen Collection.* Dir. Douglas Petrie. Writ. Douglas Petrie. Twentieth Century-Fox Home Entertainment, 2005. DVD.

"The Gift." *Buffy the Vampire Slayer: The Chosen Collection.* Dir. Joss Whedon. Writ. Joss Whedon. Twentieth Century-Fox Home Entertainment, 2005. DVD.
Golden, Christopher, and Nancy Holder. *Buffy the Vampire Slayer: The Gatekeeper Trilogy, Book Three: Sons of Entropy.* New York: Pocket Books, 1999. Print.
Gottlieb, Allie. "Buffy's Angels." *Metroactive Arts.* 26 Sept. 2002. Web. 21 Nov. 2015.
"Grave." *Buffy the Vampire Slayer: The Chosen Collection.* Dir. James A. Contner. Writ. David Fury. Twentieth Century-Fox Home Entertainment, 2005. DVD.
Greene, Richard, and Wayne Yuen. "'Why Can't We Spike Spike?' Moral Themes in *Buffy the Vampire Slayer.*" *Slayage: The Journal of the Whedon Studies Association* 1.2(2), March 2001. Web. 20 Nov. 2014.
Henze, Richard. "Deception in *Much Ado About Nothing.*" *Studies in English Literature, 1500–1900*, Volume 11, No. 2, Elizabethan and Jacobean. Houston: Rice University Press, 1971. 187–201. Print.
"The Hollow Men." *Dollhouse: The Complete Second Season.* Dir. Terrence O'Hara. Writ. Michele Fazekas, Tara Butters, and Tracy Bellomo. Twentieth Century-Fox Home Entertainment, 2010. DVD.
"The Hub." *Agents of SHIELD: The Complete First Season.* Dir. Bobby Roth. Writ. Rafe Judkins and Lauren LeFranc. Walt Disney Studios Home Entertainment, 2014. DVD.
"The Initiative." *Buffy the Vampire Slayer: The Chosen Collection.* Dir. James A. Contner. Writ. Douglas Petrie. Twentieth Century-Fox Home Entertainment, 2005. DVD.
Johnson, Kate. "A Whedonite's Guide to Characterization: Villains Are People Too." *After the Avengers: From Joss Whedon's Hottest, Newest Franchises to the Future of the Whedonverse.* Ed. Valerie Frankel. Chicago: PopMatters Media, 2015. 41–44. Print.
Jung, C.G. *The Portable Jung.* Trans. R.F.C. Hull. New York: Penguin, 1976.
Knehans, Greg. "Buffy, Betrayal, and the Ethic of Truths." *Slayage: The Journal of the Whedon Studies Association* 12.2–13.1 (40–41), Winter 2014/Spring 2015. Web. 20 Nov. 2014.
"Man on the Street." *Dollhouse: The Complete First Season.* Dir. David Straiton. Writ. Joss Whedon. Twentieth Century-Fox Home Entertainment, 2009. DVD.
"Maveth." *Agents of SHIELD.* ABC. KXTV, Sacramento. 8 Dec. 2015. Television.
McLaren, Scott. "The Evolution of Joss Whedon's Vampire Mythology and the Ontology of the Soul." *Slayage: The Journal of the Whedon Studies Association* 5.2(18), September 2005. Web. 20 Nov. 2014.
Milavec, Melissa M., and Sharon M. Kaye. "Buffy in the Buff: A Slayer's Solution to Aristotle's Love Paradox." *Buffy the Vampire Slayer and Philosophy: Fear and Trembling in Sunnydale.* Ed. James B. South. Chicago: Open Court, 2003. 173–184. Print.
Much Ado About Nothing. Dir. Joss Whedon. Writ. Joss Whedon. Roadside Attractions, 2013. DVD.
"Nothing Personal." *Agents of SHIELD: The Complete First Season.* Dir. Billy Gierhart. Writ. Paul Zbyszewski and D.J. Doyle. Walt Disney Studios Home Entertainment, 2014. DVD.
"The Only Light in the Darkness." *Agents of SHIELD: The Complete First Season.* Dir. Vincent Misiano. Writ. Monica Owusu-Breen. Walt Disney Studios Home Entertainment, 2014. DVD.
"Orpheus." *Angel: The Complete Fourth Season.* Dir. Terrence O'Hara. Writ. Mere Smith. Twentieth Century-Fox Home Entertainment, 2004. DVD.
Pollefeyt, Didier. "The Morality of Auschwitz? A Critical Confrontation with Peter Hass' Ethical Interpretation of the Holocaust." *Good and Evil After Auschwitz: Ethnical Implications for Today.* Eds. Jack Bemporad, John Pawlikowski, and Joseph Sievers. Hoboken: KTAV, 2000. 119–138. Print.
Ramachandran, Nandini. "The Big Bad Universe: Good and Evil According to Joss Whedon." *Joss Whedon: The Complete Companion: The TV Series, the Movies, the Comic Books, and More.* Ed. Mary Alice Money. London: Titan Books, 2012. 75–82. Print.
"Reunion." *Angel: The Complete Second Season.* Dir. James A. Contner. Writ. Tim Minear and Shawn Ryan. Twentieth Century-Fox Home Entertainment, 2003. DVD.
"Revelations." *Buffy the Vampire Slayer: The Chosen Collection.* Dir. James A. Contner. Writ. Douglas Petrie. Twentieth Century-Fox Home Entertainment, 2005. DVD.

Riess, Jana. *What Would Buffy Do? The Vampire Slayer as Spiritual Guide*. San Francisco: Jossey-Bass, 2004. Print.
"S.O.S." *Agents of SHIELD: The Complete Second Season*. Dir. Vincent Misiano and Billy Gierhart. Writ. Jeffrey Bell, Jed Whedon, and Maurissa Tancharoen. Walt Disney Studios Home Entertainment, 2015. DVD.
Sakal, Gregory. "No Big Win: Themes of Sacrifice, Salvation, and Redemption." *Buffy the Vampire Slayer and Philosophy: Fear and Trembling in Sunnydale*. Ed. James B. South. Chicago: Open Court, 2003. 239–253. Print.
"School Hard." *Buffy the Vampire Slayer: The Chosen Collection*. Dir. John T. Kretchmer. Writ. David Greenwalt and Joss Whedon. Twentieth Century-Fox Home Entertainment, 2005. DVD.
"Seeing Red." *Buffy the Vampire Slayer: The Chosen Collection*. Dir. Michael Gershman. Writ. Steven S. DeKnight. Twentieth Century-Fox Home Entertainment, 2005. DVD.
Selenethedaydreamingwriter. "*Agents of S.H.I.E.L.D.*: Sexual Violence and the Double Standard Within the Fandom." *Audi Alteram Partem*, 7 Nov. 2014. Web. 20 Nov. 2015.
Serenity. Dir. Joss Whedon. Writ. Joss Whedon. Universal Studios Home Entertainment, 2005. DVD.
Smirnov, Kristen. "Domestic Abuse and Gender Role Reversal in Season 6: My Letter to Mutant Enemy." *Tabula Rasa: Love, Redemption, Spike*, Web. 9 Dec. 2015.
Souza, Christopher. "Boyd Langton and the Fantasy of Trust." *Inside Joss' Dollhouse: From Alpha to Rossum*. Eds. Jane Espenson and Leah Wilson. Dallas: BenBella, 2010. 205–217. Print.
Symonds, Gwyn. "'A Little More Soul Than Is Written': James Marsters' Performance of Spike and the Ambiguity of Evil in Sunnydale." *Slayage: The Journal of the Whedon Studies Association* 4.4(16), March 2005. Web. 20 Nov. 2014.
_____. "'Solving Problems with Sharp Objects': Female Empowerment, Sex, and Violence in *Buffy the Vampire Slayer*." *The Aesthetics of Violence in Contemporary Media*. New York: Continuum, 2008. 126–149. Print.
"The Target." *Dollhouse: The Complete First Season*. Dir. Steven S. DeKnight. Writ. Steven S. DeKnight. Twentieth Century-Fox Home Entertainment, 2009. DVD.
Thomas, Kaitlin. "Can *S.H.I.E.L.D.*'s Agent Ward Be Redeemed? *Buffy* and *Angel* Say Yes!" TVwww, 29 Apr. 2014. Web. 20 Nov. 2014.
Tresca, Don. "Dancing in the Sky: The Value of Love in *Runaways*." *The Comics of Joss Whedon: Critical Essays*. Ed. Valerie Estelle Frankel. Jefferson, NC: McFarland, 2015. 133–144. Print.
"Turn, Turn, Turn." *Agents of SHIELD: The Complete First Season*. Dir. Vincent Misiano. Writ. Jed Whedon and Maurissa Tancharoen. Walt Disney Studios Home Entertainment, 2014. DVD.
"Two to Go." *Buffy the Vampire Slayer: The Chosen Collection*. Dir. Bill Norton. Writ. Douglas Petrie. Twentieth Century-Fox Home Entertainment, 2005. DVD.
"Welcome to the Hellmouth." *Buffy the Vampire Slayer: The Chosen Collection*. Dir. Charles Martin Smith. Writ. Joss Whedon. Twentieth Century-Fox Home Entertainment, 2005. DVD.
"Why We Fight." *Angel: The Complete Fifth Season*. Dir. Terrence O'Hara. Writ. Drew Goddard and Steven S. DeKnight. Twentieth Century-Fox Home Entertainment, 2005. DVD.
Wilcox, Rhonda V. "Echoes of Complicity: Reflexivity and Identity in Joss Whedon's *Dollhouse*." *Fantasy Is Not Their Purpose: Joss Whedon's Dollhouse*. Eds. Cynthea Masson and Rhonda V. Wilcox. Special issue of *Slayage: The Journal of the Whedon Studies Association* 8.2–3(30–31), Summer/Fall 2010. Web. 20 Nov. 2014.
_____. "Every Night I Save You: Spike, Sex, and Redemption." *Why Buffy Matters: The Art of* Buffy the Vampire Slayer. Ed. Rhonda V. Wilcox. London: I.B. Tauris, 2005. 79–89. Print.

The Detective, the Pastor and the Thief
Backstrom's *Negotiations of Morality*

ALISSA BURGER

The single-season Fox series *Backstrom* (2015) is based on a trilogy of Swedish crime novels by Leif G.W. Persson. Persson's Bäckström books, which include *Linda, as in The Linda Murder* (2005), *He Who Kills the Dragon* (2008), and *The True Story of Pinocchio's Nose* (2013) and are part of "The 'Nordic crime wave' that has made many Scandinavian authors familiar around the world [and] features a variety of world-weary cops with broken families" (Davis 11). According to Nordic noir author Jørn Lier Horst, protagonists within this tradition tend to share some common characteristics, including being "taciturn [and] slightly uncommunicative ... constantly embarked on an uncompromising pursuit of truth and clarity" and "primarily distinguished by their single-mindedness and their rebellion against authority in all its facets." While Persson has not yet achieved the mainstream popularity of Horst, Stieg Larsson, or Jo Nesbø, the American television adaptation of his series retains several of these characteristics, including the abrasive and morally complex titular character of Detective Everett Backstrom (played by Rainn Wilson) and a complicated family history that includes an abusive father—who also happens to be a legendary lawman—a deceased mother, and a half-brother whose existence Backstrom discovers midway through the series. Fox's *Backstrom* combines these elements with a twist on the standard police procedural, following the cases of a Portland, Oregon Special Crimes Unit, comprised of a ragtag ensemble of investigators who have been passed on from other assignments or departments, a group that Backstrom refers to with mingled admiration and derision as a "unit of losers" (Episode 1.1, "Dragon Slayer"). As his summation of his team suggests, Backstrom can be a tough man to like. He is coarse, rude, drinks and smokes to excess, has an affinity for pornography and prostitutes, and lives a generally unhealthy and self-destructive lifestyle. These habits impact his police work, and he often transgresses the boundaries of procedure and legality. He jumps to conclusions and frequently makes stereotypical snap judgments about suspects and crime scenarios, lies about being shot by a suspect, forges a warrant, plants evidence, and intimidates and physically abuses a suspect. Despite his abrasiveness, however, Backstrom gets results. He is a consummate good guy/bad guy: what he does is often wrong, misguided, or even illegal, but in doing so he solves cases and catches bad guys. He is a very successful investigator, has a great track record of finding answers and closing cases where others have failed, and his job is all-consuming; once he's on a case, he thinks of little else until he has figured it out, obsessively fixated on finding the criminal and bringing him or her to justice. However, beneath this single-minded drive, his motivation, more often than not, is simultaneously fueled by his self-serving need to gratify his enormous ego.

Backstrom isn't the only man who often occupies this moral gray area in *Backstrom*. Detective John Almond (played by Dennis Haysbert) is a pastor as well as a detective, and might be expected to be a moral compass for the wayward Backstrom. However, like Backstrom, Almond is willing to look the other way, come to Backstrom's rescue, and even cross those lines himself when necessary. Pulling Backstrom even further into the realm of illegality is his friend, roommate, and half-brother Gregory Valentine (played by Thomas Dekker). Valentine is a thief and fence for stolen property. As a result, the barge on which Backstrom and Valentine live is a uniquely liminal space: while the residence of a policeman, it is also a clearing house for stolen property, in a living arrangement and relationship that complicates the usual cop/criminal dichotomy. Backstrom, Almond, and Valentine are each multifaceted combinations of good guy and bad guy, presenting a wide range of constantly and dynamically negotiated moralities and masculinities, deconstructing and questioning the contemporary police procedural format, negotiating definitions of the hero and the antihero, and delving into the moralistic and character-driven gray areas within each man.

Law, Order and the Antihero

In recent years, antiheroes have dominated primetime, from *The Sopranos*' mafia boss Tony Soprano (HBO, 1999–2007) to serial killer *Dexter* (Showtime, 2006–2013) and *Breaking Bad*'s meth mogul Walter White (AMC, 2008–2013). These are just a few of the morally compromised men that are part of "a notable recent movement—serial narratives organised around fundamentally immoral, unsympathetic and anti-heroic characters" (Smith 30). As Murray Smith argues of these series' popularity, "There can be little doubt that there is a large audience for the morally distressed narrative, and that the sustained exploration of weak, dubious, compromised and corrupt protagonists is a major reason for the critical and popular success of these programmes" (31). These small screen antiheros also demand a more complex negotiation from their viewers than the traditional good guy/bad guy dichotomy, where right and wrong are relatively easy to distinguish. As Margrethe Bruun Vaage lays out this character trope in *The Antihero in American Television*, the antihero is

> a clearly—or even, severely—morally flawed main character whom the spectator is nonetheless encouraged to feel with, like and root for. The moral complexity of the antihero series entails that the spectator is intended to like the antihero—but through a challenging narrative also come to dislike him. The antihero series typically encourages sympathy for the antihero initially, but increasingly also questions this positive orientation with the antihero [Vaage].

The audience's conflicted response to the antihero is integral in his meaning and significance. In Smith's "Mad, Bad and Dangerous To Know: TV's Anti-Heroes," he argues that "Imagined, vicarious transgression may be one part of the equation, but these series defeat any simple response of either sympathy or antipathy, pleasure or revulsion, towards their protagonist, cultivating instead a complex and profound ambivalence" (31). While dichotomously polarized good and bad guys are designed to elicit simple audience responses—identifying with the hero, cheering the downfall of the villain—antiheroes demand a much more nuanced and actively negotiated understanding.

Traditional police procedurals have long trafficked in the good guy/bad guy dichotomy with upstanding cops and unethical villains the norm, with these series "structured as morality tales, with a heroic figure or team setting out each week to bring criminals to justice" (Nichols-Pethick 6). As Astrid Dirikx, Jan Van den Bulck, and Stephen Parmentier argue, popular culture police act as "societal moral agents" (38) and within this capacity "[s]everal studies have found that one way in which the police show that they represent society's moral values is through the procedural fairness with which they exercise their authority" (39–40). These fictional police officers and detectives are heroic because they can be trusted, motivated by a moral quest for truth and justice. However, in recent years there has been "a transition from narratives about lone cops to ensemble cast [and] from good-versus-evil to more muddied morality" (Sabin 2). Series like *The Wire* (HBO, 2002–2008) and *The Shield* (FX, 2002–2008) have offered a darker twist on the traditional police procedural. For example, *The Shield*'s Vic Mackey "refrains from nothing to keep his power in the district, and his crimes include murder—even of fellow police officers" (Vaage). As these series demonstrate, the line between good guys and bad guys is often porous and shifting, a perspective that informs many contemporary primetime television series, including *Backstrom*.

Backstrom upholds some of the characteristics of the traditional police procedural series while deviating significantly in other ways. As Jonathan Nichols-Pethick explains in *TV Cops: The Contemporary American Television Police Drama*, "Typically, the character of the police officer has been predicated on a range of heroic abilities, including physical prowess, finely-honed instinct, wisdom gained through experience, a healthy suspicion of policy bureaucracy (bordering on disdain), a sense of selflessness, a commitment to one's peers (especially one's partner), and an intimate knowledge of criminal circles" (Nichols-Pethick 68–69). Another recent development of the police procedural is the "expanded heroic action to include emotional bonding and social sensitivity" (Nichols-Pethick 69), a change that impacts not only relationships among police officers but their interactions with victims, witnesses, and suspects as well, with police fairness in part assessed based on their treatment of the public "in a dignified and polite way" (Dirikx, Van den Bulck, and Parmentier 40). Backstrom's dedication to his work is unquestionable and unparalleled. He is among the best of the best at closing cases and getting results, but these achievements are often tainted by his less than professional demeanor, disregard for police procedure, disdain for colleagues and victims, and often inappropriate interactions with both.

Detective Backstrom is, in some ways, a hero and an outstanding detective: he is relentless in tracking down criminals and is haunted by the cases he couldn't solve and the victims he failed to save; his dedication to his job borders on the obsessive, and his instincts are often spot-on. However, the way in which he goes about his investigations is often more akin to vigilantism than through officially-sanctioned processes, including collecting evidence without a warrant, falsifying warrants, planting evidence, and lying to suspects, all of which jeopardize the successful prosecution of these cases. So despite his commitment to catching bad guys, as a result of his failure to follow procedure, it often remains unseen whether these arrests will hold up or be feasibly prosecutable, and while Backstrom achieves the instant gratification of getting criminals off the street—and often more importantly to him, winning and making sure he hasn't been outsmarted by those criminals—his actual impact on the safety of Portland is uncertain. In fact, this responsibility is passed on and it is left up to his team to make sure that his cases are prosecutable. As Backstrom's partner Nicole Gravely (Genevieve Angelson) tells their fellow investigator John Almond following a questionable instance of evidence collection, "Backstrom's job is to solve the crime. It is my job to make sure that he does so in a way that can hold up in court" (Episode 1.3, "Takes One to Know One"). While Gravely to some extent excuses Backstrom's bad behavior as she takes on the responsibility of holding him accountable and handling the procedural matters that often get lost or disregarded in Backstrom's hedonistic, ego-driven obsession, the whole team is just as guilty of this as Gravely, as each of Backstrom's team members repeatedly cover for his transgressions and deviations from procedure. In the larger discussion surrounding the antihero, this faith can be problematic. As Joshua Alston argues, one "narrative problem with antiheroes is not that they are flawed but that they are flawless.... Antiheroes don't get the luxury of being wrong, and audiences are robbed of the opportunity to watch these characters deal with the consequences of their mistakes." Backstrom gets into trouble but very rarely has to take responsibility for his actions, and with his self-destructive habits, there's very little to stop him from breaking the rules or going too far, since he almost never has to face the consequences, either professionally of personally. Regardless of the lengths to which Backstrom goes, his team defends him and covers for him.

Backstrom has great instincts, a unique perspective, and is able to put himself effectively and often empathetically in the position of the criminal he is questioning with his trademark "I'm you…" identification. For example, when Backstrom identifies a serial arsonist as she's staking out the next house she plans to torch, he confronts her with his reflection that

> I'm you. I'm a girl with a secret that burns like a flame. I don't get turned on like other women. To really enjoy sex, I need candles, because it's the flame that gets me going, not the person in my bed. And when that stopped working, I needed to set bigger and better fires. And when that stopped working…. I needed to burn…. People [Episode 1.2, "Bella"].

While she dismisses this as "just a story," the photos on her camera and the evidence in her trunk, including her signature fire-starting setup, identify her as the arsonist, providing enough evidence to arrest her and try her for the previous fires. However, Backstrom's willingness to go with his instincts often leads him to knee-jerk stereotypical conclusions. He is both politically and culturally insensitive. For example, when a professional sex surrogate is found murdered, he persists in referring to the victim as a "hooker," having to be repeatedly corrected and reminded by his team members that she is, in fact, a "former prostitute" and "not a hooker" (Episode 1.10, "Love Is a Rose and You'd Better Not Pick It"). As *The New York Times*' Alessandra Stanley explains, Backstrom is an "ill-mannered misanthrope who enjoys making sexist and racist remarks … and solves crimes by insulting suspects and making huge deductive leaps based less on observation than obstinacy—he follows his biases." He makes stereotypical and racist remarks when working with Native Americans on a nearby reservation (Episode 1.7, "Enemy of My Enemy"; Episode 1.13, "Rock Bottom") and while investigating the murder of a Chinatown psychic (Episode 1.6 "Ancient, Chinese, Secret"); he seamlessly equates a local mega-church with a cult ("Takes One to Know One"), immediately assumes a victim's stripper girlfriend is a gold-digger rather than a legitimate romantic partner ("Dragon Slayer"), and struggles with pronouns and frank discussions of sex and gender when he and the team work to solve the murder of a transgender victim (Episode 1.4, "I Am a Bird Now"). While Backstrom's team members frequently call him out and correct him on his racist and sexist remarks, this criticism is compromised by the fact that Backstrom's gut reactions—often prompted just as much by his prejudices and preconceptions as by his deductive reasoning—are often proven correct or at the very least, provide a clue or lead that helps the team solve the case. Just as with his team's acceptance of and covering for Backstrom's deviation from the procedural rules, they also all too often excuse his offensive generalizations and biases, even occasionally justifying his skewed perspectives.

One suspect challenges Gravely, asking, "This guy hates women. You know that, right? Why do you help him?" to which Gravely responds, "I don't actually think he hates women more than he hates men.... He hates everybody" ("Bella"). This response is characteristic of the larger series, where the only people who consistently challenge Backstrom are women: Gravely, Chief of Police Anna Cervantes (played by Inga Cadranel), and the head of the disciplinary Civilian Oversight Committee, Amy Gazanian (played by Sarah Chalke), who also happens to be Backstrom's ex-fiancée. While these three women question Backstrom on numerous occasions, they do so rather ineffectually, their concerns are largely dismissed, and they have little choice but to either capitulate to Backstrom's methods or distance themselves from his antics and his arguably toxic influence. Gravely similarly comes to Backstrom's defense when a drug treatment facilitator challenges Backstrom about his addiction issues, retorting that "Lieutenant Backstrom is conflicted with everyone about everything" (Episode 1.9, "The Inescapable Truth"). True to the larger tradition of police procedurals, in which cops have the backs of their fellow officers, especially their partners, Gravely comes to Backstrom's defense, even when doing so glosses over or elides the real issues of bias, bigotry, and snap judgment on which she herself frequently challenges him.

While Backstrom problematically embodies some of the representational expectations outlined by Nichols-Pethick, there are others at which he fails completely. He is not physically fit and the series opens with him seeing a department-mandated physician (played by Rizwan Manji) who has to give Backstrom a clean bill of health so he can continue working, which is challenging in light of Backstrom's self-destructive lifestyle, unhealthy eating habits, cigar smoking, and alcoholism. Backstrom is helpless when confrontations with suspects turn physical, and "he is not physically brave—when confronting an armed killer intent on escape, he is more jelly-kneed than most television cops" (Stanley). Between his physical unfitness and his fear, he often has to rely on his teammate Officer Frank Moto (played by Page Kennedy), who also happens to be a mixed martial arts fighter and more than capable of protecting Backstrom from angry suspects and giving chase when necessary. Returning to the characteristics Nichols-Pethick discusses, Backstrom also lacks selflessness, thinking first and foremost of himself, his needs, and the gratification of his enormous ego. For example, in "Bella," the series' second episode, Backstrom and his team are investigating a series of arsons. While other members of Backstrom's team follow up on legitimate leads, interviewing experts and staking out the most recent arson site, Backstrom becomes obsessed with drawing a connection between the fires and a couple of the responding firefighters, who bullied Backstrom as a child. Backstrom is so invested in getting his revenge that he uses a falsified warrant to gain access to the firehouse, plants evidence on one of the firefighters, and manipulates them in an attempt to get them to turn on one another. Finally, Backstrom is less motivated by solving the arsons—though thanks to his team and Valentine, the arsonist is identified and arrested—but more by getting back a homemade kite that the firefighters stole from him decades earlier.

Another element of the typical police procedural from which Detective Backstrom deviates significantly is the expectation of fairness and trustworthiness. As Dirikx, Van den Bulck, and Parmentier argue, this expectation has a significant impact in the public's perception of the police because "if authorities treat people fairly, they communicate that they represent normative group values" (40). Fairness is determined in part by "relational aspects of procedures ... [including] trustworthiness [and] neutrality" (Dirikx, Van den Bulck, and Parmentier 40). While Backstrom only incompletely embodies the characteristics outlined by Nichols-Pethick, he falls far short of these measures of fairness. As Dirikx, Van den Bulck, and Parmintier explain, "Trustworthiness pertains to the motives of an authority in reaching decisions. An authority should care about people's needs and should not be led by self-interest. Neutrality refers to the objectivity of an authority. Authorities should make their decisions based on facts and rules and should not allow personal biases into their decisions" (40). Backstrom falls far short of this expectation and regularly puts his own perspectives, preconceptions, and personal desire ahead of the case, even when that means suspects and victims are treated unfairly. As evidenced in the "Bella" episode, the legality of the investigation is secondary to Backstrom's own personal vendetta against his childhood bullies, and he is willing to lie, steal, and manipulate evidence to get the outcome he wants. Backstrom lies about being shot by a suspect—in fact, he accidentally shot himself, just before shooting and killing an unarmed suspect—both on his official report of the incident and later, before the Civilian Oversight Committee in their further investigation ("Dragon Slayer"). Arguably, one of the biggest repercussions of Backstrom's moral flexibility is the ways in which it impacts his fellow teammates, both individually and collectively. When Backstrom shoots an unarmed suspect in the series' pilot, the first Special Crimes Unit team member on the scene is Moto, who sees Backstrom getting rid of the suspect's gun in the river. Moto is later called to testify before the Civilian Oversight Committee about the shooting and is unsure of what he should say, since he didn't witness the shooting and has questions and reservations about what actually happened. In the end, Moto lies under oath, perjuring himself to support Backstrom's series of events. When Almond confronts Moto about lying, telling Moto that in giving in to Backstrom's pressure, "He made you a bad guy," Moto responds by pointing out the greater good, that Backstrom "catches more bad guys than anyone… . Being good isn't about following the rules" ("I Am a Bird Now"). In covering for Backstrom, Moto also believes he is doing what's best for the team, standing by his partner, and showing that he can be trusted in a crisis. As he tells Almond, "Someday you're gonna wonder to yourself if I'm somebody you can count on. And now you know you can" ("I Am a Bird Now"). As long as the end justify the means, resulting in a criminal behind bars, under Backstrom's direction, there's plenty of room for moral flexibility.

The Pastor and the Thief

When a character is faced with a moral crisis, they often turn outward, looking beyond their own conflicted conscience for some sort of counsel or guidance. The image of the individual in turmoil with an angel on one shoulder and a devil on the other is a familiar trope, externalizing the moral negotiations and though processes that can go into making a tough decision or responding to a complex and complicated scenario. While Backstrom rarely cares much about what anyone else thinks of him or his actions, when he does he turns to two particular friends: detective and pastor John Almond and Gregory Valentine, Backstrom's tenant, underworld connection, and lately discovered half-brother. While elements of these characters' identities—man of God and criminal, respectively—would seem to position them easily as the angel and devil on Backstrom's shoulders, the series sidesteps this easy, dichotomous characterization, instead giving viewers two sophisticated and morally complex characters.

Almond is often the voice of reason, policy, and protocol within the Special Crimes Unit, as evidenced by his discussion with Moto that no matter the cost, lying for Backstrom was not the right call. Almond upholds this morality outside the police station as well, where he is a pastor of the small, store-front style Joy of Everlasting Light Church. Almond's police work influences his preaching and vice versa. When he and the team hit a dead end in a child abduction case, he takes time out of their fruitless search to lead his fellow team members in prayer (Episode 1.5, "Bogeyman"). However, when it comes to catching killers, Almond's faith seemingly provides him some of the same moral flexibility as his team members, though never to the point of perjuring himself under oath like Moto. For example, he goes along with Backstrom to the firehouse even though he suspects the warrant that takes them there is falsified ("Bella"), covers for Backstrom with a prosecutor when Backstrom is in the interrogation room interviewing a suspect while high on stolen Oxycodone ("The Inescapable Truth"), and shoots through a padlock that stands between them and the site of a suspected body dump ("Rock Bottom"). While Almond often chastises Backstrom and tries to hold the detective to a higher standard, he also sees the good in him. Following Backstrom's recovery of a missing child and the related solving of a missing persons cold case, Almond is giving a sermon in his church celebrating the young women's safe return. As Almond tells his gathered congregation, "We all know that evil can cloak itself in beauty. But good can also confuse us in deportment and appearance," as the images cut between Backstrom passed out alone in his barge and those who have come together in Almond's church, including the girls, their families, and the rest of the Special Crimes Unit. Morally speaking, Backstrom may not always make the right choices, as Almond reminds Moto and others at various points throughout the series, but overall, Almond sees and values the good in Backstrom.

Almond's faith in God and his dedication to doing good police working and serving his community are unshakable and absolute, in stark contrast to Backstrom's self-serving laissez faire approach to police work and life in general. Almond is unquestionably a "good man," while Backstrom dodges any easy moral categorization. However, Almond's goodness also allows him to see and appreciate more of Backstrom than meets the eye, to understand and embrace the reality that there is more than one way of being a "good man." Almond has seen Backstrom at his best and at his worst, including a self-destructive binge following a case Backstrom failed to solve ("Bogeyman"). As a result, like the other members of Backstrom's team, Almond is willing and able to bend the rules and look the other way when necessary, though only when the sin, so to speak, will lead to the salvation. Though Almond's core values are rarely compromised, his actions are at times morally ambiguous and while he never occupies the liminal space of the good guy/bad guy as completely as Backstrom, he actively questions, critiques, and negotiates these moral categorizations, whether in the police station or the pulpit.

If Almond attempts to further develop the good in Backstrom and urge him toward the moral path, Backstrom's relationship with Valentine aligns him more clearly with the criminal element of Portland. Backstrom and Valentine have a unique relationship: while on the surface they seem to be diametrically opposed—cop and criminal—the reality is much more complicated. Valentine is Backstrom's tenant, a connection they first forged because Backstrom is old friends with Valentine's mother, a prostitute he frequented; midway through the series, they also discover that they are half-brothers, further highlighting their connection and camaraderie. In their current situation, Valentine lives with and stores stolen merchandise on Backstrom's barge, a thief and fence carrying on his illegal activities within the home of a policeman. Almost inevitably, Valentine's illegal activities occasionally intersect with Backstrom's investigations, creating a conflict of interest that is raised multiple times over the course of the series. As series creator Hart Hansen explained in an interview, Valentine "works right at the edges" (qtd. in Halterman) and while his main means of income come from fencing stolen goods, he's also one of Backstrom's frequent paid informants and provides the Special Crimes Unit with connections and entries into the underworld that they would not otherwise have. For example, in "Bella," when Backstrom is trying to prove that firefighters are stealing from burned houses, he sends Valentine in an unofficial undercover capacity, paired with Detective Peter Niedermayer (played by Kristoffer Polaha) to investigate and recover the stolen goods, a task that nearly jeopardizes Valentine's livelihood, just as Valentine's criminal activity later jeopardizes some of Backstrom's cases.

Valentine is seen by Backstrom's team members as a threat to their success and even their very existence as a functioning team. Both Moto and Almond encourage Backstrom to cut ties with Valentine, and just as Backstrom's fast and loose approach to investigating often takes him outside the bounds of legality, the conflict of interest posed by Backstrom and Valentine's relationship threatens to have significant repercussions for the Special Crimes Unit and their cases. As Almond tells Backstrom of Valentine, "he's jeopardizing your career and the careers of the people you lead" (Episode 1.12, "Corkscrewed"). Moto and Almond's concerns prove founded when Backstrom assaults Dante Trippi (played by Jim Dreichel), a violent criminal who beat Valentine, an interaction which becomes even more problematic when Trippi is found dead outside of Backstrom's barge ("Corkscrewed"). While Backstrom's assault of Trippi is violent, disturbing, and clearly illegal, it also highlights a choice that structures much of Backstrom and Valentine's relationship, one which significantly transcends the good guy/bad guy dichotomy. As Nichols-Pethick argues, "'crime' is always subject to [the] tension between legal definitions and more experiential or circumstantial definitions. These competing definitions of crime are also played out in relation to ideas of 'community'—how the interpretation of crime can vary according to the dynamics of a particular group" (18). Just as Moto lied for Backstrom to protect him and the team as a whole, Backstrom and Valentine are willing to cover for one another. However, given their varying positions as cop and criminal, these multiple—and often at odds—identifications also often require them to lie *to* one another as well. This dissonance often causes tension between the two men as they question each other, though their protection of one another is unwavering. As Valentine tells Backstrom, "our relationship isn't based on truth, it's based on trust.... We tell each other what the other needs to know" ("Corkscrewed"). However, the moments when they have to leave one another in the dark prove problematic, both personally and professionally. For example, in the investigation of one of the Special Crime Unit's cases, Backstrom arrests one of Valentine's former boyfriends (Episode 1.11, "I Like to Watch"), while Valentine finds himself blackmailed by Trippi following Backstrom's assault ("Corkscrewed"). In the end, they simply have to accept the fact that they will always have secrets, some of which may endanger their personal and professional relationships, but that they will also always have one another's backs. It's a high-risk relationship, structured by situational morality and founded in part on the lies they must tell one another, but it's also simultaneously one of the most honest, transparent, and fulfilling relationships in the series.

Just as Almond's characterization challenges the easy moral categorization of the "good" pastor, Valentine fails to fit the mold of the "bad" criminal. Valentine's actions are clearly self-serving and illegal, though he also often puts that illegality to good use, drawing on his knowledge of the Portland crime scene to help Backstrom and his team solves crimes. Significantly, both of these men also accept Backstrom just as he is, neither trying to reform or remake him. Backstrom's trust in these two men is nearly absolute, though both that trust and their support of him exist in constantly negotiated gray areas, just as each of them navigate the liminal and morally complex spaces between good guy and bad guy.

Backstrom the Antihero?

Lieutenant Everett Backstrom is very good at what he does, puts a lot of bad guys behind bars, and lives to solve crimes. However, he's also a biased, self-centered, and self-destructive egomaniac. He's offensive, obnoxious, and often difficult to like. As Ashley M. Donnelly argues of antiheroic characters like Backstrom, "Though their individual acts are called into question ... each time this interrogation into the potentially villainous behavior of 'good characters' occurs, the hero ultimately is defended. The all-encompassing argument is that heroes such as these, even if they do terrible things, will give all of themselves to avenge, protect, or save the innocent, even if they themselves come close to losing their humanity" (8). While Backstrom's first imperative may be to gratify his own ego and to "win" at any cost, he does so through an entirely unique perspective, seeing things that others miss, and those victories put criminals behind bars, stopping murderers and saving lives.

Finally, it is essential to consider the position of the antihero, viewers' negotiated responses to this figure, and his often conflicted response to himself. As Vaage explains, one of the hallmarks of the antihero is the complex series of often contradictory responses he elicits from viewers. Backstrom himself echoes these same responses, a self-reflexive antihero who shifts between smug self-satisfaction and self-loathing. As Nancy DeWolf Smith explains, despite Backstrom's biases and regular mistreatment of just about everyone around him, "He probably doesn't hate others so much as he hates himself," a combination of his abusive childhood, his volatile relationship with his father, and his dark worldview. However, no matter how pessimistic Backstrom's perspective of the world, it is authentic. As he says early in the series, "The truth is always dark.... That's how you know it's the truth" ("Bella"). The series ends with a final unanswered moral question, with Backstrom himself uncertain of whether he is a hero and an antihero. In the series finale, Backstrom finally defeats his father—and the demons of his childhood—but feels conflicted about his victory and himself. Finally standing before the Alcoholics Anonymous meeting his doctor has been pushing him to attend, Backstrom is truly alone for the first time in the series. There is no one to make apologies for him or bail him out; as he talks about the people who have impacted his life, including Valentine and Special Crimes Unit team, they are shown in contrast to his raw and confessional voiceover narration. In an episode—and series—that began with Backstrom on the verge of being relieved from duty, that is a question that remains unanswered as he stands alone before a group largely made up of strangers, with the exception of his department-mandated physician. As he takes the first step in dealing with his addictions, Backstrom is heartsick, tearful, and uncertain as he looks into the surrounding faces and asks, "So now what?" ("Rock Bottom").

This question, which lingers beyond the closing scene of the series, creates a possibility for Backstrom to grow and change, to redeem himself and reclaim his life. As Smith argues of televisions' antiheroes, "The possibility of redemption, no matter how slight or remote, is thus an essential ingredient in these shows" (33). The antihero maintains the possibility of heroism in his own questioning of himself and this self-reflection creates the opportunity for his salvation, though coming to terms with the self can be a struggle, just as viewers find themselves questioning and negotiating their allegiance with their favorite small screen antiheroes. As Smith continues, "self-deception is key to the possibility of redemption ... [antiheroes] strive to dupe themselves because at some level they understand, but can't stand, what they have become" (ibid). Viewers may like Backstrom, even cheer him on in some of his most misanthropic moments, taking vicarious pleasure in his antisocial antics, but in this final scene, the viewer is positioned alongside Backstrom, reexamining not only his disregard for politeness, political correctness, and personal feelings, but looking beyond this brash exterior to the inner shadows of Backstrom's self-loathing and self-destructiveness, as well as their own responses to these antiheroic traits and the complex and conflicted hero that—just maybe—lies beneath.

Works Cited

Alston, Joshua. "Too Much of a Bad Thing." *Newsweek* 153.2 (2009): 58–59. *Academic Search Complete.* Web. 8 Dec. 2015.

Davis, J. Madison. "He Do the Police in Different Voices: The Rise of the Police Procedural." *World Literature Today* 86.1 (2012): 9–11. *Academic Search Complete.* Web. 10 Dec. 2015.

DeWolf, Nancy Smith. "The Curmudgeonly Crime Solver." *The Wall Street Journal.* Dow Jones & Company, Inc., 15 Jan. 2015 Web. 10 Dec. 2015.

Dirikx, Astrid, Jan Van den Bulck, and Stephan Parmentier. "The Police as Societal Moral Agent: 'Procedural Justice' and the Analysis of Police Fiction." *Journal of Broadcasting and Electronic Media* 56.1 (2012): 38–54. *Academic Search Complete.* Web. 8 Dec. 2015.

Donnelly, Ashley M. *Renegade Hero or Faux Rogue: The Secret Traditionalism of Television Bad Boys.* Jefferson, NC: McFarland, 2014. Print.

Halterman, Jim. "What's the Straight/Gay Relationship All About on 'Backstrom?' Creator Hart Hanson Clues Us In!" *Xfinity.* Comcast, 28 Jan. 2015. Web. 10 Dec. 2015.

Horst, Jørn Lier. "What's the Secret of *Nordic Noir*?" *Norway: The Official Site.* N.p., n.d. Web. 21 July 2016.

Nichols-Pethick, Jonathan. *TV Cops: The Contemporary American Television Police Drama.* New York: Routledge, 2012. Print.

Sabin, Roger, with Ronald Wilson, Linda Spiedel, Brian Faucette, and Ben Bethell. *Cop Shows: A Critical History of Police Dramas on Television.* Jefferson, NC: McFarland, 2015. Print.

Smith, Murray. "Mad, Bad and Dangerous to Know: TV's Anti-Heroes." *Culture & Religion Review Journal* 2014.3 (2014): 30–34. *Academic Search Complete.* Web. 8 Dec. 2015.

Stanley, Alessandra. "A Caustic Detective on the Beat: 'Backstrom' Stars Rainn Wilson as Yet Another Overbearing Genius." *The New York Times.* The New York Times Company, 21 Jan. 2015. Web. 8 Dec. 2015.

Vaage, Margrethe Bruun. *The Antihero in American Television.* New York: Taylor & Francis, 2016. Ebook.

Building and Breaking an Antihero
The Rise of Sonny Corinthos

JACINTA YANDERS

Soap operas have been a consistent part of my life from as far back as I can remember. Before I started school and later during holidays and breaks, I often found myself underfoot of my grandmother, who was a dedicated viewer of *All My Children, One Life to Live,* and *General Hospital*. She loved her "stories," and she passed that love on to me. Particularly, she seemed to have the strongest affinity for *General Hospital*. While she was certainly invested in many storylines and characters over the years, there was perhaps no character that she enjoyed quite as much as Sonny Corinthos. Sonny is the resident mob boss of *General Hospital*'s fictional town of Port Charles, New York. When my grandmother talked about Sonny, she'd say, "There's my man." She almost unilaterally disliked characters that opposed Sonny on the show. To my grandmother, the very flawed Sonny, with all of his charm and his notorious dimples, was a character worth rooting for. She was certainly not alone in this belief. Sonny Corinthos has been an extremely popular character over the years, especially in the 1990s. But how did his popularity come to be? How did a character that started on *General Hospital* as a sleazy amoral strip club owner grow into not only a beloved character, but also a character that is arguably most central to the show in its present form?

To understand the evolution of Sonny Corinthos in the 1990s, it is useful to understand a bit of soap history. Different decades have seen different types of stories take the lead; thus, the prevalence of some story types has a lot to do with conventions of the time period. In the 1980s, for example, stories that specifically focused on romance and adventure were extremely popular. While such stories cropped up to some extent on many soap operas,

General Hospital in particular is a soap that was especially well known for such storylines.

With stories that paired adventure with romance, viewers witnessed the rise of the soap opera "supercouple." On the most basic level, a supercouple is a pairing that is found to be extremely popular amongst fans. And while there were such couples before the 1980s, the term is particularly associated with that decade. According to Elana Levine:

> These soap supercouples were romantic pairings that combined stories of love lost and found with action-oriented intrigue and adventure. Their fantasy oriented tales were a marked turn away from the narrative and representational tendencies of the US daytime soaps of the 1970s, which tended toward more realism, less glamour and closer attention to social issues [22].

Understanding the types of stories being told, those that emphasized action and exploration, is crucial to the understanding the types of characters that were being written in these years. And understanding these storylines and character types further helps to elucidate the dramatic shift in storytelling that would take place when Sonny arrived on the scene. While his storylines have certainly been marked by what many would classify as action and adventure, and while he's certainly had his fair share of romances, the ways in which those particular Sonny-centric stories developed were quite different from the somewhat lighter style of the 1980s soaps. The shift in the types of stories being told is often reflective of the cultural climate. After all, the 1980s were supposed to be a decade of relative ease compared to the decades of significant struggle that had preceded it. And even if that ideal wasn't reflected in everyday life, it seems fairly likely that viewers may have looked to soaps as a means of escape from "real life." Those fantasies and adventures gave them the opportunity to see a different life, even if it wasn't one that they could experience.

While viewers were living vicariously through the characters, they had the opportunity to experience intense love stories in addition to adventure. It's true that love stories have always been part of the soap opera genre, but the love stories of the supercouples in the 1980s were more tightly focused. As Levine says about the 1980s, "the stories that took up the most airtime, and that inspired the biggest viewer response, were about characters falling in love, facing obstacles along the way" (29). In fact, this emphasis on romance coincided with what was perhaps ABC's most significant tagline, "Love in the Afternoon." This tagline was used in numerous popular promos in the 1980s, and it still comes up in contemporary soap opera conversations as representative of what soaps can and should be. This is perhaps especially relevant given that even though, as I mentioned previously, Sonny Corinthos has had several romances over the course of his time on *General Hospital*, his other dominant, primarily mob-centric, storylines have often been criticized in recent years for taking up too much space on the show and not allowing other romances and storylines to share in the spotlight.

Given soap operas' significant interest in fantasy in the 1980s, it may come as no surprise to find that many, if not all, of these stories were made up of characters that could generally be categorized as either heroes or villains. The villains were often obvious. They were the ones trying to destroy the world or a town or a couple. They were violent and/or devious. They sought power over others. They left devastation in their wake. In summary, they were the bad guys, and they weren't necessarily written to be rooted for. On the other hand, the heroes were supposed to be where viewers showed support. These were the good guys. They had the love stories and the adventures. They were the ones who worked hard to triumph over everything that the villains threw in their paths.

Admittedly, this does not mean that the heroes were perfect people in the most conventional sense. There was often tension between the two halves of the supercouples. Generally, there was some notion that the male in these pairings was a "bad boy" who would potentially corrupt and/or ruin the woman in the pairing. The bad boy label does not imply that these men were villains. They were still typically characterized as good, helpful people, but perhaps they didn't have a steady job, they rode a motorcycle, or they engaged in some other behavior that, while not necessarily being outright malicious, was frowned upon for some reason. This long history of the good "bad boy" may have, in particular ways, set up expectations that ultimately allowed Sonny Corinthos to flourish. As I will discuss shortly, his character was never meant to become a focal point for the show, but perhaps part of the reason that viewers did latch on to the character has to do with the fact they'd been primed to see these youthful miscreants as redeemable love interests.

Though soaps' 1980s bad boys may have started out in the margins when their storylines began, they were typically brought into the fold and remolded into more upstanding gentlemen over time. This clarified that they were the ones to root for as they battled adversaries. Of course, they may have still colored outside the lines on occasion. As Levine says, they "remained edgy, willing to skirt the letter of the law" (32). Naturally, it makes sense for these characters to have maintained a certain degree of wildness so that they could more easily have adventures. And they needed to be somewhat willing to bend or even break the rules to stand toe-to-toe with the villains.

However, not every non-villain character that was part of these storylines in the 1980s can easily be classified as a true, or even willing, hero. Instead, he might be more widely recognized as an antihero. The antiheroes were people who either had done terrible things in the past or people who continued to do terrible things, perhaps for understandable reasons, while also being endearing and while exhibiting signs of possible redemption. As many have recently noted, contemporary television shows have become enamored with antiheroes. And while it would likely be impossible to trace the impetus of this influx back to any one particular factor, it seems probable that at least one significant step in the process was *General Hospital*'s abundantly popular character, Luke Spencer.

Like Sonny Corinthos, Lucas Lorenzo Spencer was never intended to be in Port Charles long-term. However, he remained a significant part of the show, on and off, for several decades. One of the most significant contributing factors to this lengthy run would have to be the fact that Luke and his better half Laura were arguably the supercouple of all supercouples. Luke and Laura garnered a degree of fame and prominence that was beyond difficult to match. Their wedding in 1981 attracted thirty million viewers, which was something that was highly unprecedented for daytime television (Wolf).

But despite this supercouple status and despite engaging in many acts of heroism, it's markedly difficult to categorize Luke as one of the good guys. His character was brought on the show in an attempt to assist his sister, Bobbie, in her quest to break up the golden couple of Laura and Scotty. Of course, this sort of mischief could be overcome by the heroes of the 1980s. However, Luke's next act, raping Laura, could not be managed as easily. In fact, one could argue that the show, in choosing to describe the rape as a seduction for the better part of two decades, simply chose not to manage it at all. But this rape is at the very core of who Luke Spencer is. His self-deprecation and feelings of unworthiness are tied to the rape just as much as they're tied to his abusive childhood. Luke may have helped to save the world from being frozen during the Ice Princess storyline, but the title of hero is not one that he would find most comfortable. For many years, Luke was involved, in some capacity or another, with the mob. He wasn't inclined toward a career in law enforcement like some of the other supercouple men of the 1980s were. In fact, he found the fact that his son, Lucky, became a cop to be rather disheartening. Luke simply wanted his freedom, and he wanted freedom for his family. His doing good deeds were more of a matter of happenstance rather than a personal moral imperative. Yet, despite his problematic nature, in the same way that my grandmother supported Sonny, many viewers rooted for Luke. They believed in his potential for love and redemption, and they saw beyond the flaws.

In the years since Luke's arrival, there have been other soap opera characters that share many of the same antihero traits. They often had similarly tragic childhoods, and they may have even committed similarly heinous acts, such as rape. They also committed acts of heroism on a situational basis, and they struggled to stay on the right side of the law. They carried the weight of their pasts with them, and even though they sought love, they often found ways, consciously or unconsciously, to sabotage the love that they were given. They were not the cartoon villains, but they couldn't easily be called heroes either. They didn't just occasionally color outside of the lines like many of the 1980s bad boys, but instead, they often lived there. And if one were to rank which of these characters has embraced this style most effectively, it's very likely that the name Sonny Corinthos would find its way to the top of the list.

When the 1990s began, the tone of *General Hospital* began to change as well. In some regards, the show returned to realism in a manner similar to what had been prevalent in the 1970s. While there were still grand adventures to be had, they were fewer and farther between. Storylines took on a darker and/or more realistic tone and characters became more difficult to classify as heroes or villains. The overt fantasy element of the 1980s began to fade away. To some extent, this could be considered a return to the core of soap operas. After all, soap operas have "a special sense of feeling 'true to life' ... the drama of soap opera is the drama of everyday life" (Barbatsis and Guy 59). Admittedly, soap operas certainly heighten the intensity of such drama, but it does seem to make sense that part of the thing that continues to bring viewers back is this semi-realistic reflection of our world. In the 1990s, issues of violence, disease, and drugs were often in the news. It seems logical then that the show, in its attempts to be reflective, would take a darker turn. This darker, grittier Port Charles was the perfect place for Sonny Corinthos to make his entrance.

Michael "Sonny" Corinthos, Jr., first appeared on our televisions screens in 1993. With a name that alluded to *The Godfather* fairly obviously, it would have been shocking if Sonny turned out to be anything other than a gangster. However, the character was not initially at the top of the crime food chain. After all, given that the character wasn't meant to stick around, it wouldn't have made sense to give him such a status. In fact, Sonny's portrayer, Maurice Benard, had no interest in joining the show long term. He had acted previously on a different soap and was not inclined to do so again. But, like everyone else, he needed to work. When *General Hospital* approached him with two possible characters he could play, Damian Smith with a two-year contract or Sonny Corinthos with six-month contract, Benard chose the latter (Mulcahy, Jr.,). Ironically, the Damian Smith character only lasted for a few years, but Sonny Corinthos has been a mainstay of *General Hospital* for roughly two decades.

The earliest episodes involving Sonny Corinthos provide a rough sketch of who Sonny was, while giving the slightest glimmers into what he could become. He ran a strip club and seemed to have little problem with using young women for personal gain. In fact, this was the root of Sonny's initial storyline with the young and beautiful Karen Wexler. Karen was, for all intents and purposes, a good girl, but she was struggling to cope with the presence of trauma in her life, which left her open to the influences of Sonny. He encouraged her to strip in his club, and when Karen had difficulty balancing her schoolwork, stripping, and the general struggles of her life, Sonny provided her with pills. This is done, ostensibly, under the guise of helping Karen, but what it more directly does is help to further develop a situation in which Sonny can take advantage of Karen. At times, Sonny seems to be genuinely interested in Karen, but the show makes clear that his motivations are primarily self-serving. This makes sense given the type of character that he was initially and the limited amount of time he was supposed to be in Port Charles. It's not, after all, unheard of, for a short-term character to be brought on as a plot device to advance the narrative of a more foundational character like Karen.

However, Sonny Corinthos was given room to grow. The show began to capitalize on the charisma that Benard's portrayal of Sonny seemed to have in spades. This particular asset is one of the most common when it comes to television antiheroes. Jason Mittell notes that such characteristics cause viewers to "overlook their hideousness, creating a sense of charm and verve that makes the time spent with them enjoyable, despite their moral shortcomings and unpleasant behaviors" (76). If Sonny had been continuously portrayed as a charmless predator, it's unlikely that he would have found such solid footing on the show.

But creating an antihero that fans will invest in requires more than charismatic performances. The writing of the character has to be structured in such a way that viewers will buy into the character's worthiness. A change in the show's writers happened around the same time as Sonny's arrival in Port Charles. The new writers that inherited this fairly undeveloped character were not particularly pleased with the character's trajectory at the time. In fact, head writer Claire Labine has said that she initially found Sonny "really detestable" and that she "hated the stripper story" (Jacobs). As such, Sonny was re-calibrated and developed into a more nuanced and viable character, a complexity that the audience received favorably. One early analysis describes the character of Sonny as "enigmatic" and "realistic and sexy" (Williams). The same analysis also notes the tensions between the best and the worst elements of Sonny, his selfishness and his compassion, the latter of which is primarily revealed via his close friendship with the younger, struggling, ultimately AIDS-stricken Stone Cates. Viewers constantly had to wonder what element of Sonny was most dominant. Was he simply a victim of the circumstances of his life who had potential for good or was he a more malevolent figure?

This tension would continue to be explored in many ways throughout Sonny's early storylines. A particular intervention that the writers made on behalf of the character was to reveal more and more about his difficult childhood. It was a childhood in which both Sonny and his mother were brutally abused by Sonny's stepfather Deke, who happened to be a cop. The effects that this abuse had on Sonny would continue to reverberate over the years. While one might argue that it wasn't completely necessary for Sonny to have an abusive childhood for fans to support him, it was, at the very least, important for him to have some relevant backstory. The audience needed to know more about whom he really was and about the experiences and choices that led him to his life of crime. Not only does his backstory help viewers understand the character, but it also helps them to form a connection with the character. As Mittell says, "The more we know about a character through revelations of backstory, relationships and interior thoughts, the more likely we will come to regard them as an ally in our journey through the storyworld" (76). It's unclear to what extent such an in-depth exploration of Sonny's childhood was initially planned. Though the first contract was short, the beginning of this revelation did occur during Sonny's interactions with Karen. Perhaps then it was only conceived of as a part of the plot, a tool with which Sonny could further draw Karen in, given that she had also experienced abuse.

However, the story was expounded on with much more depth in Sonny's subsequent relationship with the character Brenda Barrett. Brenda had been on the show for some time before Sonny's arrival, and had established herself on the canvas. It was not immediately obvious that Sonny and Brenda would be placed into each other's orbit. However, once they were paired up, there was no going back. Claire Labine describes their connection as "magical" and notes that there was serious investment on the part of the writers in Sonny and Brenda as a couple (Jacobs). Together, Sonny and Brenda formed a tremendously popular couple. They had the occasional adventure, a fair amount of romance, and plenty of physical attraction. However, like Luke, Sonny was not classifiable as a hero. His acts of heroism were similarly happenstance, and there was even less of a chance that Sonny, given his chosen career path and his association with his abusive, police officer stepfather, would find his way to the right side of the law.

Despite being part of a beloved pairing, Sonny's path on the show was not easy by any means. Following the closure of his strip club, Sonny was more prominently entrenched in the mob. After some time, Sonny even took over the entire territory of Port Charles. As these moves occurred, Sonny struggled to maintain his relationship with Brenda. For a long time, she didn't know the extent of his criminal lifestyle, which put her at significant risk. Though Sonny did seem to have the best intentions with Brenda, his career decisions led him down a path of danger and destruction. The show foregrounded these contradictory elements of Sonny's personality often. For example, viewers saw him being a dedicated caretaker for Stone. At the same time, viewers also saw Sonny break laws and exhibit violent rage. When Brenda wore a wire as a means of finding out who Sonny really was and this was revealed to Sonny, he treated her extremely harshly. Despite his love for her, he was not easily able to understand or forgive what he saw as a betrayal, even if Brenda's decision logically made sense.

Notwithstanding these contradictions, or perhaps because of them, 1990s Sonny Corinthos maintained a significant level of public approval. He may have been a gangster, but he was more than that. He may have broken laws, but he also took care of people. He wasn't always victorious, but he worked hard to get and maintain what he had. In a sense, Sonny's struggles and vacillations likely made him easier to connect to than the heroes of the 1980s. As Jenn Bishop says, "There was a sense that he wanted to do the right thing, yet wasn't always sure what that was, something many of us can relate to..." ("Why General Hospital's Sonny Needs to Face Retribution"). Sonny was not perfect by any means, but he was striving for the best possible life. And while he may not have always played by the rules, he did not generally seem to intentionally be hurting others. His trials and triumphs made him accessible. Furthermore, Sonny was emotionally vulnerable, particularly with the people he cared about. This is, of course, why Brenda's "betrayal" hurt him so much. He'd opened himself up to her, and she had, in his opinion, violated his trust. Dealing with the aftermath of Stone's AIDS diagnosis similarly showcased Sonny's vulnerabilities. Here was a man who Sonny had come to love as a brother, who was dying, and there was nothing that Sonny, with all of his power and control, could do about it. Vulnerabilities such as these played a major part creating a complex and endearing character.

Despite Sonny's best intentions, he continued to struggle with darkness, and his decisions put those he cared about at risk. Following his relationship with Brenda, Sonny married Lily Rivera at her mobster father's behest. Though their relationship was more platonic than anything else, Lily did ultimately become pregnant, and Sonny committed to a life with his wife and child. The possibility of fatherhood brought out the best in Sonny, perhaps because he "wanted a child of his own to give the love that he was never given" (Bishop). Ultimately, that happy ending was ripped away when Lily fell victim to a car bomb planted by her father. Sonny had been the intended target, but Lily and their unborn child were the casualties. This loss devastated Sonny in ways that had been previously unimagined and ultimately may have been responsible for Sonny's even more fervent grasp for control in the future. This desire for control remains one of his most consistent characteristics today.

Though it wasn't immediate, Sonny and Brenda did reunite following Lily's death. But Sonny was unable to ignore the reality of the danger his career choices placed Brenda in. He was also aware of the fact that the mob isn't a job that one just casually walks away from. It wasn't as though he could simply become a bank teller or a teacher. Seeing no way in which he could easily attain a new life with Brenda and knowing that continuing to be with her would put her in danger, Sonny chose to walk away from her on their wedding day. He made the decision to put Brenda first, in his own way. He broke her heart, so that she could let him go and have a safe life. This was, perhaps, the peak of Sonny doing what some would consider the wrong thing for the right reasons. His decision to walk away crushed Brenda, but it was inarguably the most successful move Sonny could make to keep her safe. His decision coincided with Maurice Benard's departure from the show, but he would not be gone for long.

When Benard returned to *General Hospital*, almost a year after his departure, the show was beginning to take a slightly different path. A different set of writers than those who had written Sonny in the early 1990s were in charge, and storylines became more tightly focused on smaller groups of people. There were less storylines that revolved around the hospital, and more and more storylines that began to focus on the mob. Notably, this shift toward more of a mob focus happened at approximately the same time as *The Sopranos* became one of the most critically acclaimed shows on television (Scodari 114). The Quartermaine family, which had been prominent on the show for several years, began to slowly dwindle, and the Corinthos family began to take a more central role. Throughout the early 2000s in particular, Sonny's extended family dominated numerous storylines. And as such, yet another shift in Sonny's character began to take place.

As Sonny amassed more airtime and centrality, he seemed to simultaneously lose many of the vulnerabilities that endeared audiences to him throughout the 1990s. A narrative began to take shape where Sonny and his cohorts could never lose, at least not in any lasting way. It didn't matter whether their opponents were other mobsters or the police; the Corinthos family was almost always victorious and generally written in a way that suggested that they were on the right side of the dispute. And where Sonny had been quite significantly influenced by the death of his wife and child in the 1990s, the many losses that he endured in the 2000s did not have lasting impact. Where 1990s Sonny would have likely taken more responsibility for his involvement in these losses and would have made more measured attempts to lead a better life, 2000s Sonny rarely operates as such. Instead, he clings to his power, denigrates his "enemies," and continues to put his loved ones in danger while simultaneously paying lip service to the notion that he can keep them safe. This particular change in characterization runs counter to typical expectations for antihero characters. Whether explicitly stated or not, they're expected to always be seeking redemption. And while they may "dupe themselves because at some level they understand, but can't stand, what they are or have become," they certainly aren't supposed to truly buy into the lie as one might argue Sonny Corinthos has (Smith 33).

Additionally, and importantly, Sonny Corinthos was diagnosed with bipolar disorder. This occurred, in part, due to the fact that Maurice Benard himself has bipolar disorder and the writers wanted to incorporate this into the storyline. But even before this mental illness was explicitly written in the show, viewers had witnessed Sonny having multiple emotional breakdowns. Before the diagnosis, these breakdowns were not typically described as possible evidence of a mental illness. Similarly, the many instances in which Sonny has engaged in verbal tirades, thrown things, and so on over the years were often glossed over by other characters. After the on-screen diagnosis, one might expect that a more emotionally stable Sonny Corinthos would have emerged. However, Sonny still exhibits much of the same behavior today. His violent temper and verbal assaults, in particular, are still present, suggesting perhaps that those qualities have less to do with Sonny's mental health and more to do with who he truly is.

Though the contemporary characterization of Sonny is one in which his vulnerabilities and true struggle are not at the forefront, that does not mean that character does not still occasionally perform acts of heroism. Fairly recently, Sonny saved a boatload of people from dying by jumping into the harbor with a bomb. There's an extent to which such moves could still be read as attempts to redeem the character, but one might argue that the egregious horrific acts that Sonny Corinthos has committed in recent memory typically outweigh whatever good he might do. For example, he murdered his adopted son's biological father after promising not to do so. The details of that particular situation were complicated, but his actions clarified, more than ever, that Sonny Corinthos is capable of causing catastrophic damage. While rival mobsters have been killed, locked up, and/or forced out of the business because they're seemingly "worse" than the Corinthos family for whatever reasons, Sonny remains in power. It is not at all unusual to see television antiheroes contrasted "with more explicitly villainous and unsympathetic characters to highlight the anti-hero's more redeeming qualities" (Mittell 75). However, the degree of regularity with which this happens in Sonny's storylines may be exceeding the acceptable norm. As for the real "good guys," the Port Charles Police Department semi-regularly tries to bring charges against Sonny, but almost always comes up short. And even Sonny's oldest son, Dante, who is a detective, has difficulty siding against his father for very long. Sonny has transitioned from someone who would be knocked down and have to work to get back up into someone who is always on top, regardless of the circumstances.

Perhaps in another possible attempt at finding redemption for Sonny, the writers have recently paralyzed the mob boss. This is the first time that

Sonny has been in a truly vulnerable situation in several years, and the show has focused on exploring how he rebels against his obvious need to depend on others now. Though he's still in charge of his criminal empire, Sonny shows signs of feeling weak because he's been confined to a wheelchair. To him, his power is slipping away, and there's little that he can do to change that, which frustrates him endlessly. Simultaneously, Sonny's youngest son, Morgan, has also recently been diagnosed with bipolar disorder. Sonny has had to witness his son's erratic behavior firsthand, and he struggles with finding the best way to help Morgan cope, knowing the problem can't be easily fixed with money or an exertion of strength. Theoretically, these two events could potentially lead to some much needed character growth for Sonny Corinthos. Whether that actually happens, as of the time of the writing of this essay, remains to be seen.

Ultimately, the character of Sonny Corinthos has experienced a complex arc over the course of his twenty years as part of *General Hospital*. Recent years have seen his actions have becoming less and less like the endearing antihero. For what it's worth, when Maurice Benard was asked about how he sees Sonny, he said, "I see Sonny as a gangster with a good heart. He loves his family, he loves his kids, but he'll take you down if you cross him. He does bad things (at times) for good reasons. He's human" (White-Nobles). This conceptualization seems to be very much in line with the antihero Sonny, but is it truly representative of who Sonny is today? Has Sonny gone too far into the depths of violence and unchecked power? Has he moved too far from the grey space into one that lacks rootability? Is he still an antihero? Or has he, perhaps, become the villain?

Works Cited

Barbatsis, Gretchen, and Yvette Guy. "Analyzing Meaning in Form: Soap Opera's Compositional Construction of 'Realness.'" *Journal of Broadcasting & Electronic Media* 35.1 (1991) 59–74. Print.
Bishop, Jenn. "Why General Hospital's Sonny Needs to Face Retribution." *TVSource Magazine*. SoSource Media, 9 Nov. 2014. Web. 17 Nov. 2015.
Jacobs, Damon L. "Soap's Hope: The Claire Labine Interview, Part Four." *We Love Soaps*. N.p., 6 Nov. 2009. Web. 17 Nov. 2015.
Levine, Elana. "Love in the Afternoon: The Supercouple of 1980s US Daytime Soap Opera." *Critical Studies in Television* 9.2 (2014): 20–38. Web. 17 Nov. 2015.
Mittell, Jason. "Lengthy Interactions with Hideous Men: Walter White and the Serial Poetics of Television Anti-Heroes." *Storytelling in the Media Convergence Age : Exploring Screen Narratives*. Hampshire: Palgrave Macmillan, 2014. Web. 8 Apr. 2016.
Mulcahy, Kevin, Jr. "Exclusive: Maurice Benard Interview, Part 1." *We Love Soaps*. N.p., 23 Dec. 2010. Web. 17 Nov. 2015.
Scodari, Christine. "Of Soap Operas, Space Operas, and Television's Rocky Romance with the Feminine Form." *The Survival of Soap Opera: Transformations for a New Media Era*. Jackson: University Press of Mississippi, 2011. Print.
Smith, Murray. "Mad, Bad and Dangerous to Know: TV's Anti-Heroes." *Culture & Religion Review Journal* 2014.3 (2014): 30–34. Print.
White-Nobles, Omar. "Interview with a Mobster: 'General Hospital's' Maurice Benard on the

Changes in Sonny's Life and GH's Future." *TVSource Magazine*. SoSource Media, 9 Aug. 2012. Web. 17 Nov. 2015.

Williams, Carol Traynor. "Soap Opera Men in the '90s." *Journal of Popular Film & Television* 22.3 (1994): 126. Print.

Wolf, Buck. "Luke and Laura: Still the Ultimate TV Wedding." *ABC News*. ABC, 15 Nov. 2006. Web. 17 Nov. 2015.

You Can't Go Home Again
The Multiple Heroisms of Sergeant Nicholas Brody

LLOYD ISAAC VAYO

For the first three seasons of the Showtime drama *Homeland*, Sergeant Nicholas Brody, a former Al Qaeda captive who returns to the United States to face the machinations of political opportunists and the suspicions of CIA agent Carrie Mathison, takes center stage. From the first episode, Brody is framed as a hero: though held captive and subjected to torture, Brody (hypothetically) stands strong during his confinement, eventually being freed by U.S. forces and greeted with pageantry upon his arrival stateside. Yet, for viewers and, increasingly, for Mathison and Agency cohort Saul Berenson, that heroism comes under suspicion, as evidence mounts suggesting that Brody was turned by Al Qaeda higher up Abu Nazir.

However, Brody's heroism is not so easily discounted, or so simply constructed. Rather, his heroism is threefold. The first iteration, the hyperbolic hero soldier archetype, is in full flower upon Brody's return, as he is lauded for his service to and noble sacrifice for his country. This is the most straightforward and stereotypical of the heroisms that Brody embodies, and is seen as the natural role for him to embody and for others to identify within him. The second iteration is coincident with the first, an alternate heroism in which the evil deeds of Vice President William Walden, who frames the strike that kills Abu Nazir's son Issa (who Brody befriended in captivity) and his schoolmates as a successful operation against Al Qaeda militants, are avenged by the eventual success of Brody's plot against him. Brody's assassination of Walden represents a privileging of pure justice over patriotism, avenging Issa's death and putting aside the first heroism for an intensely personal and idealistic form. Following an interregnum in which Brody operates as a double agent, playing the CIA and Abu Nazir against each other, the third iteration appears at the end of season three. After Brody flees to Venezuela following his identification as a suspect in the bombing of CIA headquarters, Berenson recruits him to (successfully) assassinate Danesh Akbari, head of the Iranian Revolutionary Guard. By virtue of this assassination, Brody is able to redeem his perceived (though not actual) participation in the Langley bombing, enacting a conflicted heroism that paradoxically cleanses him by getting his hands dirty in pursuit of an abstract notion of justice. Though Brody's cover is blown and he is eventually hanged, he dies a hero, three times over: lost on the battlefield and redeemed on return, a man without a country who dies for that country, and a soldier who does good by doing evil.

Before examining Brody's own unique brand(s) of heroism, it will be useful to ground the analysis within more traditional notions of heroism as a means of highlighting the exceptional version(s) embodied by Brody in *Homeland*. Taking the work of Carl Jung (in *Man and His Symbols*) and Joseph Campbell (in *The Hero with a Thousand Faces*) as a starting point for a survey of the more traditional hero archetype, the basics of that archetype may be identified. For Jung, "[t]he universal hero myth ... always refers to a powerful man or god-man who vanquishes evil ... and who liberates his people from destruction and death" (Jung 68), and Brody fulfills that standard in his actions against Walden and Akbari. Jung maintains that "hero myths vary enormously in detail, but the more closely one examines them the more one sees that structurally they are very similar" (101), and indeed this is the case for most manifestations of the myth, making Brody's unique version doubly so in its complication of that standard structure. Elsewhere, Jung anatomizes four distinct cycles of the hero myth (via Dr. Paul Radin's *Hero Cycles of the Winnebago*): the trickster, the hare, the red horn, and the twins, and it is the latter that proves of most interest and accuracy in the case of Brody. Jung describes the twins thusly: "Though the Twins are said to be sons of the Sun, they are essentially human and together constitute a single person. Originally united in the mother's womb, they were forced apart at birth. Yet they belong together, and it is necessary—though exceedingly difficult—to reunite them" (106). For Brody, this twin-ness manifests in his heroic dualism; he is at once the stereotypical, hyperbolic patriotic hero and the "terrorist," an agent of both vengeance and redemption, and the task of uniting those heroisms is his greatest preoccupation, though not an entirely successful one.

The manner in which Brody proves himself an exception to that rule may be elucidated with the help of Campbell's influential study. After moving through creation and mythology, Campbell turns his attention to the place of the hero in human history, and in the course of locating the hero, he identifies a cycle of transformation that the hero undergoes. Though Brody finds himself moving through several of these stages of transformation, he garbles the order laid out by Campbell, in the process defining himself as an exceptional case. Rather than concluding as "the saint or ascetic, the world renouncer" (Campbell 354), Brody begins there via his time in captivity and return as the patriotic hero. While in captivity, he finds himself "dwelling in solitude, eating but little, controlling the speech, body, and mind, ever engaged in meditation and concentration, and cultivating freedom from passion" in the words of the *Bhagavad Gita* (qtd. in Campbell 354), and he brings that sainthood back with him stateside upon his return. Brody then backtracks to the hero-as-warrior stage, "the champion not of things become but of things becoming… [who] [w]ith a gesture as simple as the pressing of a button … annihilates the impressive configuration [of the tyrant]" (337). Brody makes himself the champion of Walden, who is potentially in the process of becoming President and achieving even greater power, and intends to do so by pushing the detonator button on his suicide vest. Brody then moves to the stage of hero-as-redeemer, enduring "the period of desolation caused by a moral fault on the part of man" so as to slay the tyrant (352), doing so with the knowledge that "[t]he hero of yesterday becomes the tyrant of tomorrow, unless he crucifies *himself* today" (353). On the path to his final action against Akbari, Brody must first endure a dark night of the soul, questioning his motives and plumbing the depths of the complicated nature of his heroism, before realizing his mission in an act of self-sacrifice that preserves him from the blemishes associated with CIA-affiliated black ops. Brody's halting, wrongly ordered progression through the stages of heroic transformation underlines his unique status as the complex hero of the post–9/11 American landscape. His jumbled, fragmentary transformation mimics the bodily fragmentation produced by his suicide plot against Walden and by the drone-launched Hellfire missile used on Issa, as well as the ideological fragmentation engendered by the 9/11 attacks themselves and the ensuing response. As Brody puts the pieces of his life back together, he simultaneously reassembles the psyche of post–9/11 America, his tripartite heroism suggesting that the path ahead is complex at best, and none too clear.

Brody's first iteration of heroism, the hyperbolic soldier archetype, is present most distinctly shortly after his return from purported captivity, more specifically in the episodes "Pilot" and "Semper I" (episodes one and four of season one), which cast him as a uniquely valuable heroic figure. After being flown stateside and greeted by his family, somewhat awkwardly in the case of his son Chris, who barely remembers him, Brody arrives at his home, an unassuming one-story ranch house in a non-descript suburb. Once there, Brody finds a welcoming party composed of press and well-wishers clustered around his driveway. Already experiencing culture shock due to the swift transition from captivity to domesticity, Brody can barely process what he is seeing, but his daughter Dana, oftentimes wise beyond her years, concludes, "We're famous" ("Pilot"). Dana not only grasps the larger implications of the assembled people, but also suggests that Brody's heroism is so substantial that it pulls his entire family into its orbit. Substantial welcoming parties are not typically extended to just any returning soldier, giving further evidence of Brody's unique status as a hero, above and beyond the normal use of that term for veterans. Mike, a fellow Marine and intimate companion of Brody's wife, Jessica, in Brody's absence and assumed death, adds to Brody's archetypal heroism. After Brody gives a speech to a group headed out on deployment, Mike remarks that "they're Marines and they're looking at you like you're a rock star" ("Semper I"). Mike's words endow Brody with an even greater status as a hero to heroes, one who even transcends basic heroism to take on a broader popular appeal with an altogether different cultural cachet. In this moment, Brody's saintly status becomes apparent, and he functions as perhaps the best of the good guys in the eyes of the average American. However, it is not only for the public at large and for fellow Marines that Brody attains such hero status.

If it is not impressive enough for Marine Nicholas Brody to be regarded as an idyllic version of the soldier hero by average Americans and fellow servicemen and servicewomen, Brody's fan club also extends to the intelligence community and the White House, demonstrating the reach of his particular brand of heroism. While congratulating his team on the operation that frees Brody from captivity, David Estes, Director of the CIA's Counterterrorism Center, tells them "because of you, an American hero is coming home" ("Pilot"). Estes' comments at once garner the team something approaching heroism (though the necessary anonymity of the mission renders that an impossibility) while asserting Brody's more impactful form as not only a veteran, but a veteran who has survived eight years in captivity and lived to tell the tale. So great is Brody's heroism at this stage that, upon his arrival in Washington, he is greeted by Vice President Walden in a one-on-one meeting, a gesture extended only to exceptional manifestations of the hero soldier archetype. In the aftermath of that meeting, while Brody is being presented to fellow Marines in an official ceremony, Walden asserts that "on behalf of the President of the United States and a grateful nation, it is my privilege to welcome you home" ("Pilot"). Walden simultaneously invokes the highest office in the land and functions as its agent, while also showing deference to Brody's status as being, at least in that moment, superior to his own, a figure from on high. Brody has ascended to the pinnacle of American power, being lauded by its military and commander-in-chief (by proxy), reflecting the breadth and depth of his influence and the enduring nature of his heroism beyond the flash-in-the-pan interest sometimes given to lesser archetypes.

Though Brody's heroism is more far reaching than that of a momentary media darling, it is safe to say that, in the immediate days and weeks after his return, he is experiencing something of a moment, and is a particularly desirable commodity during that time. Military brass, witnessing the raucous reception that Brody receives at the send-off for the soon to be deployed Marines, as well as his artful handling of that audience, note that "ever since that Lawrence O'Donnell interview it's been like this everywhere he goes ... he's been blogged, tweeted about, offered book deals" ("Semper I"). The O'Donnell interview, arranged to reintroduce Brody to a grateful nation through its most holy of mediums, television, kicks off a surge of interest in Brody. The interview also makes him relatable by bringing him directly into the homes of millions (some distance from the monastic cloister of captivity), and those millions crave more, in the form of further press coverage formal and informal, as well as the promise of a book-length treatment to come. Saul Berenson, CIA officer and all around canny judge of the vagaries of Beltway maneuverings, gently understates the point when he concludes that "there's a lot of political heat on the good sergeant" ("Semper I"). This observation suggests that Brody's reception is of particular notice, given that it catches the eye of a grizzled career government man who has seen many a hero come and go in his day. Brody is indeed a hot commodity, so much so that his version of the hero soldier archetype, so tried and true as to become almost mundane in many cases, still has currency in his case, demonstrating that Brody's heroism is unique enough to break from the stale frame typically accorded to a returning veteran. Brody is in many ways the answer to prayers, not only those of his family hoping for his return, but also for the CIA and the Democratic Party.

Brody-as-heroic Marine is at once captivating for the nation at large as well as all too convenient for a CIA short on feel-good stories during the broader War on Terror and a Democratic Party looking to preserve power in Washington. Estes, upon hearing that a nervous Brody is vomiting in the bathroom of the private jet bringing him to his stateside debut in Washington, states that "I need him smiling and waving like a hero like it's the fucking Macy's Day parade" ("Pilot"). Estes' concerned comments underline the necessity of a happy, healthy, and heroic Brody to uphold the image of an effectual War on Terror. The "like a hero" both hints at the constructed nature of heroism in general, to which Brody is not at all immune, as well as the stereotypical aspects of that heroism that the public comes to expect. To wit: Brody is a hero regardless of what happens after his return, having been tempered in the fire of sacrifice, but to be an exceptional, memorable hero, he must fulfill those expectations and then some, and achieve full sainthood. While watching the coverage of Brody's debut later that evening at a bar, Carrie's companion perceptively predicts that "someone somewhere is looking at him right now and thinking Congressman, Senator" ("Pilot"), and indeed operatives in the Democratic Party have just that in mind for Brody. With all that he brings to the table, Brody may serve as an injection of youthful vigor into a stagnant party, as well as giving some hawkish credibility during a time of war, all of which leads Elizabeth Gaines, a higher level party aide who works closely with Vice President Walden, to take interest in Brody. She tells him that "your story has inspired a nation in need of inspiration" ("Semper I"), and Brody proves inspirational to the party as well. Brody offers a way forward that, in its potential for increased proximity to Walden, provides Brody with a means of advancing his second heroism, as an avenging angel against the vice president.

Brody's second iteration of heroism, the avenging angel archetype, is most decidedly present in the conclusion to season one, specifically in the episodes "Crossfire" and "Marine One" (episodes nine and twelve [the season finale]), where his desire to exact revenge on Walden for previous war crimes comes into full flower. "Marine One" begins with Brody's martyrdom tape, filmed in advance of his suicide operation against Walden and assembled higher-ups from his national security team. In the course of that brief, but dense, rhetorical exercise, Brody is careful to answer accusations that might cast him as a villain should his plan come to fruition, a defense that also complicates notions of potential antihero status. At the outset, he notes that "[b]y the time you've watched this, you'll have read a lot of things about me" ("Marine One"), a statement that functions doubly as a gesture towards his preceding heroism as a formerly captive soldier, a reputation that remains in place for all but Carrie and Saul in advance of his action, as well as towards the infamy that will follow his demise and that builds upon the existing well of interest in him as a public figure. Moving beyond that more general charge, Brody points to two specific readings of the fallout of his captivity: "People will say I was broken, I was brainwashed" ("Marine One"), and artfully turns them aside. To be broken would suggest a weakness not in keeping with hero status, and whether as a soldier or a fulcrum of vengeance, Brody does the breaking. To be brainwashed would imply that Brody is not acting of his own accord, and is not in control of or responsible for his own actions, a mindset not in keeping with the hero-as-responsible actor reading given to the soldier and the avenger. Having refuted those charges, Brody's next task in the tape is to establish the particulars of his new heroism as avenging angel, as warrior for justice, which he does quite ably.

During his first heroic iteration, Brody's identity as a Marine proves central to public perception of his status, and in his second iteration, being a Marine proves equally important, though at a more personal level. Pointing to a long legacy for that identity within his family, Brody asserts that "[p]eople will say that I was turned into a terrorist ... what I am is a Marine, like my father before me and his father before him" ("Marine One"), establishing credibility in the process. Next, Brody lays out the logic behind his strike against Walden, stating that "as a Marine, I swore an oath to defend the United States of America against enemies both foreign and domestic. My action today is against such domestic enemies" ("Marine One"). This framing preempts his potential stigmatization as an un–American terrorist, the worst of the bad guys in post 9/11 America, and reframes him as a true American hero, unlike Walden. Brody literalizes his claims against Walden, branding the Vice President and his cohorts as "liars and war criminals, responsible for atrocities they were never held accountable for" ("Marine One"), and calling for "justice for 82 children whose deaths were never acknowledged, and whose murder is a stain on the soul of this nation" ("Marine One"). In this moment, Brody casts himself as the defender of all that is near and dear to the heart of American identity, while expelling the Vice President from that vision, one a hero, the other at best an antihero, if not an outright villain. Though Brody references a drone strike against a school, a strike that is unknown to the public at large due to the careful (but imperfect) cover-up efforts of Walden, and one that may yet be subject to plausible denial, his broader sense that the strike is one crime among many moves Walden beyond the realm of unfortunate mistake into the realm of deliberate malfeasance, a fact that is borne out by the details of the strike itself.

Under pressure from the curious, if occasionally fevered, mind of Carrie Mathison, Saul Berenson is moved to dig deeper into the potential lull in Abu Nazir's activity surrounding the period of the heretofore unknown drone strike. What his investigation reveals is wholly consistent with Brody's reading of Walden and his team, rendering Brody's act less maleficent and more heroic in the process, and casting Walden as the villain to Brody's warrior hero. On surveillance tape drawn from cameras trained on the decision room as the process of approving the strike is underway, Walden coolly asserts that "if Abu Nazir is taking refuge among children, he's putting them at risk, not us" ("Marine One"), and concludes that "the potential collateral damage falls within current matrix parameters" ("Marine One"). As per Brody's understanding, Walden's actions in this moment constitute a clear war crime, as he is knowingly targeting civilians and deeming that targeting not only unavoidable, but even acceptable. Walden approves the strike in terms so antiseptic that Saul is moved to remark, "Good God. Somebody actually came up with that language" ("Marine One"). In the earlier episode "Crossfire," which traces the origins of Brody's stake in the drone strike, Walden calls "images of 83 dead children ... false, created by the terrorists for propaganda purposes" in a public statement following the strike ("Crossfire"), a denial that forestalls knowledge of the strike until Saul's backchannel excavations reveal a heavily redacted report with a telltale filing number. After that report is discovered, Walden, along with Estes, who was in the room when the strike was approved, panics, and nervously says, "I thought that paper trail was burned" ("Marine One"), another action consistent with his branding as a war criminal by Brody. Saul's digging in this instance renders Brody's actions against Walden, which could quite easily be perceived as treasonous, at least understandable, if not valorous. Just why Brody takes such an interest in this particular strike is as yet unclear, though a look at his time in captivity with Abu Nazir reveals the true import of that strike.

While being held first by Iraqi loyalists, and then by Abu Nazir, Brody exists in close proximity to those commonly branded as "terrorists" and, in that proximity, gains a new perspective on Abu Nazir and his youngest son, Issa, that generates the motivating force behind his suicide plot against Walden. After being plucked from Al Qaeda torture chambers, Abu Nazir charges Brody with teaching his son Issa English, allowing him to interact freely with and develop a close bond with Issa throughout their teacher-student (and nearly father-son) relationship. Once Issa is among the dead in the Walden-approved drone strike, Brody's rage first gains a broader focus as he and Nazir watch Walden's public statement on the strike, after which Nazir remarks "and they call us terrorists" ("Crossfire"). Brody's larger anger at Issa's cruel fate soon takes on a more narrow direction, moving from the somewhat inchoate "they" to the more specific Walden. Nazir, once Brody's tormenter, becomes something of a partner in grief, a shift facilitated in part by Nazir's claim that "we only begin as enemies because that's what others told us to be" ("Crossfire"). Through Issa's death, Brody, adrift in a hopeless situation as a captive, likely believed dead after so many years, finds direction in his renewed status as a Marine, sworn to protect the U.S. against its enemies. Also, as a father, he was bound to protect those under his charge, a group expanded beyond Jessica, Dana, and Chris to include young Issa. Though he wavers somewhat on the path to his attempted suicide attack against Walden, Nazir's call for him to "never lose sight of what you committed to do in Issa's name" drives him on ("Crossfire"), pushing him through the doubt and enabling him to eventually kill Walden after the suicide plot fails. What follows in season two, Brody's framing as the bomber of CIA headquarters in Langley, primes him for the redemption that marks his third iteration as a returning, redemptive hero of sorts.

After briefly holding public office due in part to the legacy of Elizabeth Gaines, who dies during Brody's failed attack on Walden, Brody is rendered useful to Abu Nazir's larger plotting against the U.S. by serving as the vehicle for delivering a massive blow to the CIA in the form of a bombing of their headquarters that kills hundreds. Though he is a patsy in this instance, his ensuing vilification and flight from the country with Carrie's help positions him once more as a villain in need of redemption, and an arch bad guy in the meantime. One stop on his escape route from the U.S. is Caracas, Venezuela, and while there, a man cannily diagnoses his position in the world, calling him "a cockroach, unkillable, bringing misery wherever I go" ("The Star"). It is this third heroism, Brody as a prodigal son, that constitutes the conclusion of season three in the episodes "One Last Thing," "Big Man in Tehran," and "The Star" (episodes nine, eleven, and twelve [the season finale]). At this point, the man's appraisal is not far off; Brody has bottomed out, losing his family, his country and, perhaps most painfully, his identity as a Marine, and it is that last identity that Saul plays upon to recruit Brody for a final operation aimed at assassinating Danesh Akbari, head of the Iranian Revolutionary Guard. After extricating him from Caracas and helping him detox from an addiction to heroin, Saul offers Brody "a chance to be a Marine again, the man you were, before they broke you" ("One Last Thing"). This offer, a chance for Brody to take up the mantle of warrior hero once more, this time as a means of personal redemption, proves enticing enough to Brody that, with a little persuasion from Carrie, he accepts. Via Saul, Brody is given an opportunity to regain his lost heroism, albeit less in the form of reverence (the hero soldier) or vengeance (the avenging angel) than as a form of atonement for past sins.

That Brody feels himself in need of redemption at this point is beyond question, as the pathetic figure that he represents is a far cry from the stiff-backed man in uniform that returns at the outset of season one. Brody instead exists as a ruined man in sore need of something honorable to put next to his name. The only means of achieving that redemption is through the very combination of targeted violence and service to the state that has characterized the previous iterations of heroism, placing him at the complex intersection between heroism and antiheroism. Once a Marine killing foreign enemies, then a Marine killing domestic enemies, Brody once more turns his attention to the foreign as part of a plot to depose Akbari and replace him with someone more amicable to U.S. interests. While attempting to convince Brody to take part in the operation against Akbari, Carrie poses the question "What do you want, Brody? What have you always wanted? A chance at redemption?" ("One Last Thing"). Indeed, redemption has been an ongoing point of interest for Brody since Issa's death, first as a means of redeeming Issa's blood at the expense of Walden's, and then as a means of redeeming his own at the expense of the selfsame. The operation against Akbari offers just such an opportunity, and Brody seizes it, weathering unforeseen obstacles to the plot that require him to feign allegiance to the Iranian regime, and finally gaining the unattended audience with Akbari that he requires to carry out the assassination. Brody passes off his visit as one meant to inform on the U.S. asset meant to usurp Akbari's power, Majid Javadi, and to confess his own role in that plot; indeed, Brody comforts Akbari by saying that "I came here to redeem myself" ("Big Man in Tehran"), which he does, though not as Akbari may have thought. Standing over Akbari's lifeless body, Brody has seemingly redeemed himself for his past transgressions, exemplifying his final heroism and turning from bad guy to good guy, though as is often true in his case, things are not that simple.

At the moment of Akbari's death, after Brody has knocked him unconscious with a heavy ashtray and then smothered him, one might expect something closer to relief, if not joy, at having washed away previous sins and having redeemed himself in his own eyes, if not those of the world. However, Brody's lack of clear conviction before and after the assassination, only briefly forestalled by his unflinching performance of the assassination itself, reflect a more tortured sense of his mission and its effectuality in a broader moral sense. Prior to the assassination, while still operating under the guise of being (in Iranian eyes) the heroic bomber of Langley and merely seeking asylum, Brody is presented to Abu Nazir's widow, Nassrin, as a means of vetting the legitimacy of his claim for asylum, and their conversation demonstrates Brody's conflicted state. Brody describes himself as "a soldier back from war" ("Big Man in Tehran"). After Nassrin deems the Langley attack as a victory, Brody answers that "it's hard to see it like that. I lost so much" ("Big Man in Tehran") and describes his goal in Iran quite simply: "I want to stop running" ("Big Man in Tehran"). At this point, Brody is indeed a soldier back from war, returning from his battle to reassemble himself for one final duty, and he is attempting to stop running from that past not only to face it, but to redeem it. Yet, Brody himself does not seem so sure, almost willing himself to believe in the redemptive nature of his impending acts while speaking indirectly of them. Upon completing the assassination and fleeing to a safe house with Carrie, Brody concludes that "I'm not a Marine anymore. I haven't been for some time" ("The Star"), and responds to Carrie's question "Why did you agree to do this in the first place?" by saying "that is becoming less and less fucking clear" ("The Star"). Brody's inability to locate a coherent motive for his actions in this moment underlines his conflicted status as hero, as Brody's comments leave open a possible reading of his participation as craven bloodlust rather than salvational beneficence. Redemption is present in the assassination, though seeing it as such requires some mental gymnastics not only from Brody, but also from the world at large.

To the extent that redemption is present in Brody's assassination of Akbari, to the degree that he does successfully return as a heroic prodigal son, redeeming himself and vanquishing a dangerous enemy, it is seemingly limited. Brody is already a villain, his mission is covert and unacknowledged, and his demise is less than celebratory, swinging from a noose at the end of an Iranian crane, with only Brody himself and Carrie understanding the intent of his actions. Yet, it is through the eyes of that single onlooker, Carrie, who Brody asks not to attend his execution but who cannot stay away, that Brody's prodigality gains a broader audience, as shrewdly noted by Javadi once the preparations for Brody's execution are underway: "what you wanted, which was for everyone to see in him what you see ... everyone sees him through your eyes now" ("The Star"). At first, no one could see Brody as Carrie did, as less a hero soldier than a sleeper agent of Al Qaeda, as less a mad bomber than a man on a justified mission, as less a drug-addled wreck than a Marine once more. Now, in death, the world may see him as she does, as a laudable man who will go to any lengths to do what is right and to protect those near and dear to him, and in that sense a true, if flawed, hero. Indeed, after Brody's death, at a ceremony commemorating those killed in the Langley bombing, Carrie contends that "Nicholas Brody deserves a star" on the wall of those killed while serving the CIA ("The Star"), and though Director Andrew Lockhart denies her request, his description of those commemorated fits Brody quite aptly: "each cherished colleague remains a constant source of inspiration and courage. They all heard the same call to duty, and they all answered it without hesitation. They are our heroes. They are America's heroes, and that's how we'll remember them" ("The Star"). Brody's only memory may lie in Carrie's hastily drawn star surreptitiously scribbled in marker among the approved ones, but his heroic, redemptive return casts a longer shadow in the legacy of improved relations with Iran.

For the first three seasons of *Homeland*, Sergeant Nicholas Brody is the show's main figure, its beating heart, and its moral center (however conflicted it may be in any given moment), and his threefold rendering stands as the perfect embodiment of the often complex nature of heroism. The three iterations occur in descending order of legibility, the readymade hero soldier giving way to the more nuanced righteous assassin, which in turn gives way to an even subtler assassin whose targeting is more transparently benevolent (versus attacking a man one heartbeat from the presidency) while remaining morally ambivalent at best. These iterations are not mutually exclusive either, as Brody, once a heroic soldier returning from captivity, remains so, acting as a Marine and never fully making it back from his time with Abu Nazir, reacting with righteous fury at those who commit injustices against the weak and the innocent, and attempting to redeem himself at every turn, as a husband, a father, a citizen, and a soldier. He is neither an antihero nor against the notion of heroism as such; rather, Brody deepens conceptualizations of what a hero looks like and what acts may be considered heroic, offering a flawed performance of the heroic soldier archetype, a well-intentioned and justifiably inspired Vice Presidential assassin, and a potentially incomplete redemptive actor. Brody may not be the simple, easily understood and embraced hero that America wants in the uncertain world of the War on Terror, but he is the complex, difficult to parse but interpretively rich multiple hero that America needs.

Works Cited

"Big Man in Tehran." *Homeland*. Showtime. 8 Dec. 2013. Television.
Campbell, Joseph. *The Hero with a Thousand Faces*. Princeton: Princeton University Press, 1973. Print.
"Crossfire." *Homeland*. Showtime. 27 Nov. 2011. Television.
Jung, Carl, ed. *Man and His Symbols*. New York: Laurel, 1968. Print.
"Marine One." *Homeland*. Showtime. 18 Dec. 2011. Television.
"One Last Thing." *Homeland*. Showtime. 24 Nov. 2013. Television.
"Pilot." *Homeland*. Showtime. 2 Oct. 2011. Television.
"Semper I." *Homeland*. Showtime. 23 Oct. 2011. Television.
"The Star." *Homeland*. Showtime. 15 Dec. 2013. Television.

Good Bad Boys and the Women Who Love Them
Romantic Triangulation and the Ideal of Conformist Assimilation in The Vampire Diaries and True Blood

ANA G. GAL

"Tell [Sookie] I was born the night she found me. Because of her I went to my true death knowing what it means to love. Tell her thank you"
—Eric Northman, *True Blood* (2011)

In *Visual Consumption,* Jonathan Schroeder notes that the advancement of information technology—advertising and marketing in particular—has radically impacted the apparatus of consumerism in the Western space by positioning the image at the center of the economy (4). Especially since the 1980s, the act of consumption has become even less about the product and more about the ability of the product to perform a certain image or to project a consumerist ideal. Nowhere is this connection between the consumer's ideals and how specific products are marketed and consumed more noticeable than in CW's *The Vampire Diaries* (ongoing) and HBO's *True Blood* (finalized in August 2014)—two television shows that have had at least seven seasons and collectively aired over two hundred episodes since 2008. By examining the market forces that have blurred the dividing lines between heroes and antiheroes in contemporary popular media, in this essay I argue that despite its promise to challenge dominant values, the confused representation of "good" versus "bad" in these two television shows ultimately serves to legitimize and promote the ideal of conformist assimilation. More specifically, I consider how both shows use female-centered romantic triangles not only to negotiate two distinct brands of masculinity—one "good/traditional" (or socially acceptable) and one "bad/alternate" (or transgressive), but also to domesticate the "bad boy" through his obligatory adherence to heterosexuality and monogamy. Within this romantic triangle, the "bad" vampire initially rejects assimilation and embraces an ideology of domination, which entails a celebration of his "badness" that frequently degenerates into gruesome acts, such as stalking, rape, mind control, sexual deviance, and carnage. However, as I suggest in this essay, the audience, especially female, is constantly manipulated into rooting for the vampire-antihero and romanticizing his "bad boy" edge, mostly because once afforded a pivotal position within the central romantic narrative, he inevitably enters a process of rehabilitation that places him under an inevitable assimilationist pressure and that forces him to self-discipline.

Both *The Vampire Diaries* and *True Blood*, before being adapted for television, were published in a large series of novels by L.J. Smith and Charlaine Harris, respectively. Apart from functioning in installments, these franchises have supplemented profits by creating a large inventory of merchandise for fans and by organizing conventions where fan communities can celebrate their favorite characters. Each franchise has a considerable online presence, materialized in official and fan-created sites as well as social networks that often extend to an interest (if not obsession) with the real people involved in the creative act. In addition, as Nicholas Abercrombie and Brian Longhurst discuss in their study of media consumption, the acting conventions employed in contemporary television shows have switched from presentational modes (used in the theater, where the audience is enabled to experience an immediate reality and directly interacts with the performers) to representational modes (where the audience has access to a mediated experience and where the actors blur the borderline that separates the characters from the real persons performing the role) (59). This is most obvious in an account narrated on *Lopez Tonight* by Paul Wesley, the actor starring as Stefan Salvatore, the guilt-ridden vampire with a gory past in *The Vampire Diaries*:

> I was in a bar in Atlanta and one girl asked me to bite her. I get those requests often. I think I was a little drunk and I was like, "You know what? I'm gonna do it this time. I'm gonna bite her." I think she was expecting a really sexy bite on the neck, but I felt uncomfortable so I sort of just really reluctantly grabbed her arm and I kind of drooled on it a little bit. She was oddly repulsed. We kept oddly looking at each other for the rest of the night. It's awful. It's not like it is on TV [qtd. in Faller].

What Wesley's narrative implies, even if it was offered tongue-in-cheek, is that viewers are not simply content to watch the show and buy merchandise, but they also want to participate, or live out their fantasies. The nature of television shows and the promotional mechanisms used by producers have mutated not only the consumers' experience of being a participant but also how they perceive the dynamic between fantasy and reality, a dynamic alluding to a Baudrillardian hyperreality. That is, even though fans know that vampires are not real, they are so invested in the products they consume that they want them to be real.

Another practice that has fundamentally altered not only the market but also the transactions between author(s) and consumers is that contemporary viewers can make substantial interventions into the creative process by collaborating with the writers via social networks; for instance, Julie Plec, co-executive producer of *The Vampire Diaries*, has frequently corresponded with fans on Twitter in an attempt to both adhere to the "new" rules of the consumer market and find more engaging forms of expression, especially since networks today use a rather subjective rating system to measure the success of a television show. Thus, the popularity of a show is more often than not contingent upon a carefully crafted product that appeals, first and foremost, to its viewers' desires. Similar to advertising, television shows seem to be designed with the viewer's psychology in mind and employing at least two processes—identification and internalization, which Anne M. Brumbaugh defines as key processes that occur when the target audience feels a profound connection with a character or situation, not only due to shared physical appearance and moral values but also due to similar ethnoracial affiliations (68–69).

The protagonists Elena Gilbert from *The Vampire Diaries* and Sookie Stackhouse from *True Blood*, for instance, are depicted for the most part as the girl next door-type characters, despite the revelation that they do actually possess some supernatural capabilities. This emphasis on "common" femininity can be explained through the lenses of identification and internalization, in that the positioning of a female character with ordinary physical features at the center of the romance seems to target two distinct outcomes: on the one hand, this strategy encourages the female audience to undergo identification with the said character; on the other hand, it also proliferates a feminine ideal by juxtaposing the female character's physical ordinariness with extraordinary non-physical qualities that turn her into an asset for the opposite sex. Not surprisingly, both Elena and Sookie discover that they descend from a female lineage that is quite irresistible to male vampires: Elena is a doppelganger and descendant of Amara, the first immortal woman, and Sookie a human-fairy hybrid. In addition, both heroines are constantly praised and envied, often at the expense of other female characters, for their over-generosity and compassion—a "gift" that renders them very desirable in the eyes of the male characters that populate both the human world and the supernatural one. Thus, what is most significant about this particular strategy of depicting "common" female characters is that it is meant not only to reflect but also to mold how female viewers conceptualize gender roles as well as how gender and sex identity is shaped and exhibited in romantic interactions.

Both *The Vampire Diaries* and *True Blood* are primarily marketed as a romantic ménage-à-trois to a white, middle-class, heterosexual female audience by featuring a full-blown romantic love triangle centered on the female protagonist and two male aspirants to her affections, each male wooer falling within a specific typology: the self-loathing "good" vampire with suicidal tendencies, who constantly self-disciplines and defends the values upheld by the heroine, and the "bad" vampire who initially embraces his vampiric cravings and refuses to adhere to any rules. In *The Vampire Diaries*, Stefan and Damon Salvatore are brothers, whose interactions with each other and their love interest, Elena, construct the two distinct versions of "manhood" that often collide in hostile ways. The audience learns from flashbacks throughout season one that this rivalry is originally rooted in paternal favoritism. Raised in a pre–Civil War small town by Giuseppe Salvatore, an authoritarian slave owner, human Stefan and Damon are initially depicted striving for their father's affection and approval. Stefan is the "good" brother/son who conscientiously abides by the paternal/societal norm and possesses an exacerbated sense of morality and responsibility, while Damon is the "dark," older brother/son who rejects the father's authority, dishonors the family's good reputation by deserting from the Confederate army, and cultivates an ethic of sensual self-gratification. Depicted throughout the show as a man of rigid conscience and constantly burdened with guilt to the point of self-loathing and extreme dejection, Stefan pursues the self-inflicted task of redeeming himself for his transgressions as "The Ripper" (a cyclical, bloodthirsty killer leaving decapitated victims behind) not only by adopting an almost masochistic "vegetarian" diet, but also, as Ewan Kirkland remarks, by apologizing and "making amends for the inhumanity of [his] kind" (107). Unlike his brother, Damon initially embraces an ideology of hedonism, at the expense of non-vampires. Unapologetic for his "badness" and determined to sabotage Stefan's efforts to assimilate into the human culture of Mystic Falls, throughout the first season in particular he practices a kind of "vampire pride" that often degenerates into gruesome killings of innocent human beings, primarily his brother's human entourage. However, Damon's ideology shifts quite substantially as a result of his transactions with Elena. At first, he is excluded from the triangle and constantly shamed publicly for his transgressive behaviors that clash with Elena's moral conventions. Once afforded a central romantic position, he not only enters a process of systematic scrutiny, but is also placed under a significant assimilationist pressure. Even though he initially resists domestication by asserting an identity that clearly sets him apart from his brother's, he ends up developing status anxiety, which interestingly, has less to do with his desire to fit in and more with his fear of rejection or of being penalized by Elena for his moral ambiguity and excessive violations. By the end of season four, despite Damon's protests that "I am not Stefan, [so] how about you stop trying to turn me into him" ("Disturbing Behavior"), his identity has been drastically restructured, standardized even, to fit certain expectations (both the human characters' and the audience's). As Damon confesses to Elena in the episode "Founder's Day": "I came to this town wanting to destroy it. Tonight, I found myself wanting to protect it. How does that happen?" Thus, despite his initial edge, Damon ends up being nothing more than a reformed vampire with a "bad boy" image, a fake "bad boy" if you will, that serves only to enable the audience to experience his "dark" side vicariously and temporarily, without actually having to become it.

In *True Blood,* the male vampires Bill Compton and Eric Northman follow a similar polarization in their wooing of Sookie. Like Stefan, Bill Compton is nicknamed "Mr. Mainstream" for adopting a restrictive blood diet by either feeding solely on willing blood donors or drinking synthetic blood, despite its inability to provide a satisfactory alternative for vampiric sustenance. His desire to blend in, which presupposes a mandatory renunciation of the "bad" parts of his vampiric nature, places him at odds with those vampires who do not have an ethical issue with performing their predatory identities and who see "passing" as a human being as yet another masquerade that would force them to maintain hierarchical differences that construct the human race as the "true blood." As Randall Kennedy emphasizes, (racial) passing, a performance that is ultimately contingent upon the concealment of one's "real" heritage, is just "a deception that enables a person to adopt certain roles or identities from which he would be barred by prevailing social standards in the absence of his misleading conduct" (28). Despite the fact that Bill polices himself and the vampires who refuse to abide by human standards, the beneficiaries of his assimilationist practices insist that he be kept under constant surveillance.

Throughout the first four seasons of the show, Bill is subjected to constant discrimination and scapegoating on the basis of his vampiric affiliations; whenever murders are committed in Bon Temps, even though there is no direct correlation between Bill and the victims, he is systematically interrogated by the sheriff, who views his attempts to blend in the community with extreme suspicion. Thus, perhaps not surprisingly, as a result of constant marginalization, Bill's yearning to belong takes an interesting trajectory: disillusioned with the failure of vampire mainstreaming, he comes to the realization that his peripheral condition is inescapable, thereby embracing an extremist religious ideology that is primarily founded upon the belief in a vampire apocalypse. By the end of the fifth season, Bill plunges into barbarism and madness by staging large scale acts of terrorism, which culminate in the mass bombing of synthetic blood factories all over the United States. Social discrimination and "ethnic" profiling push him not only to adopt a violent reactive ethnicity but also to appropriate the "scare tactics" of his oppressors in order to destabilize the current power structure. This disruption, if not complete reversal, of the most sympathetic vampire's domestication is temporary, however, in that ultimately the female protagonist manages to rehabilitate him by invoking his "good" nature and former assimilationist value system, which leads him into committing the ultimate sacrifice out of excessive guilt; that is, he begs Sookie to give him the "true death." Unlike Bill, Eric, the "bad boy" of the series, fiercely resists naturalization and engages in patterns of behavior that violate human morality. During the first three seasons, even though Eric protects Sookie from those vampires who want to appropriate her body and blood, she constantly rejects and dehumanizes him for his refusal to surrender his "difference" in favor of a tamer one. In fact, it is solely in season four, when Eric's memories are temporarily erased by a vengeful witch, and thus his vampiric heritage, history, and affiliations forgotten, that Sookie grants him access to her body. In the process, however, he is inevitably reformed and his identity "rewritten" to accommodate her values and need for a romantic hero that can work toward assimilation.

Despite the producers' claims that *The Vampire Diaries* and *True Blood* are meant as discourses of female empowerment, an exploration of the male-female relationships that occur in the shows exposes that they both employ the romantic triangle primarily to assess and negotiate masculinity. That is, even though these television shows seem to propose romantic entanglements that operate primarily as female-centered triangles, which could potentially be seen as allowing the heroines to acquire some sort of agency, they serve in actuality to promote traditional masculinity. In *Masculinities in Theory*, Reeser, building on Sedgwick's observations about male bonding, argues that romantic triangles become merely a pretext to explore "a kind of masculine rivalry" that "is not simply based on a desire to defeat or to vanquish the rival, or to kill him off," but rather to signal the presence of a male desire "to emulate, to identify with, or to be like [his rival]." This desire to emulate, however, as Reeser explains, is intimately linked with a homophobic threat that the two men involved in the triangle attempt to escape (57). Therefore, triangulation, in which the female element functions both as an object of desire and a mediator, serves to stabilize masculinity and convey the two men's "mimetic desire" in socially acceptable ways. Simply put, "by loving a woman, man X defuses the threat of homoerotic desire by displacing the entire question of desire onto his and his rival's relationship with a woman" (61).

In *True Blood*, the erotic triangle of Sookie, Bill, and Eric reduces her body to an object of exchange and turns it into a catalyst for male bonding. The two male protagonists' relationship is initially strained by a competition for dominance. In the vampire world, entering a relationship with a human being is similar to entering a business transaction, in which the human body is a commodity to be appropriated and exchanged, in that it is owned by the vampire almost exclusively. In fact, other vampires are forbidden by vampire law to attempt a transfer of a human body without the owner's consent. More importantly, violating this contract can have serious consequences for the vampire engaging in illicit exchanges. The relationship between Bill and Eric is constantly governed by this vampire law. Bill, despite relying on it to ensure his ownership of Sookie's body and affections, is constantly jealous of Eric's attempts to destabilize the triangle in his favor, primarily due to his potential to successfully fulfill the role of romantic hero. As I argued previously, at first glance, Bill and Eric are situated at opposite ends of the binary scale, the former embodying the "good" vampire who embraces mainstreaming at the expense of his alterity and the latter the "bad boy" who posits himself against naturalization. However, as it becomes evident, especially in later seasons, the two vampires develop a kind of co-dependency, in that what seems to result from their initial hostility and interest in the same female body is an interesting homosociality that will end up prevailing over their romantic rivalry. Also, what renders this camaraderie possible is Sookie's presence as an intermediary, whose safety becomes both men's absolute priority. Even though Bill resents Eric, he remains open to the possibility of male cooperation. For instance, when confronted with common enemies, such as Queen Sophie-Anne and Russell Edgington—both attempting to appropriate Sookie as an asset—Bill and Eric suspend their rivalry, even if temporarily, to achieve a common goal, to ensure the safety of their object of desire. In many ways, this male dynamic seems to steal the spotlight that the romantic triangle enjoys in the original texts (i.e., Charlaine Harris's book series).

Similarly, in *The Vampire Diaries*, the relationship between Stefan and Damon is primarily defined by their constant exchange of women. Several love triangles, with the two brothers competing for the same woman, emerge in the show at different times. Initially, as long as their father retains the power of the household, the divide between Stefan and Damon is dramatically suppressed; that is, confronted with their father's rigid disciplinarianism and lack of empathy, the two brothers join in fraternal solidarity. This sibling dynamic is destabilized when Katherine Pierce, a vampire masquerading as a chaste Southern belle, insinuates herself in the Salvatore mansion, seduces, and turns both brothers into vampires. The original rivalry, prompted by the constant need to please the father, is apparently displaced by a sex-linked jealousy that only enhances the two brothers' initial differences. However, even though Katherine seems to assume a more active role by controlling the sexual dynamic of the triangle, she cannot escape her role as a mediator, in that Stefan and Damon seem to be able to negotiate their original rivalry rooted in paternal acceptance or rejection solely by exchanging a mutually desirable woman. This becomes primarily evident when they, like Bill and Eric, succeed in maintaining a sustainable collaborative relationship, despite their opposite personalities and resentment. For instance, when their father ascertains Katherine's true identity and captures her, Stefan and Damon suspend their rivalry and devise a plan to free her, which proves unsuccessful as they both end up dead at the hands of their father. History repeats itself later in their lives, in that the two brothers compete for another woman's affection, Elena who, like Katherine, will end up fulfilling a similar role— one that serves to mediate between the two brothers and to facilitate their reconciliation. In addition, the transactions between Stefan and Damon, similar to those established between Bill and Eric, despite developing solely as a result of triangulation, take precedence over the heterosexual romance.

An important aspect of this romantic triangulation is heterosexuality, mainly because it serves to a great extent to normalize the bad boy's/bad vampire's body. That is, the plots of the two shows create a space for both heroes and antiheroes to engage in behaviors and practices that inevitably lead to a reinforcement of mainstream values, heteronormativity in particular. In *The Vampire Diaries*, for instance, when he "turns off" his humanity and becomes "The Ripper," Stefan seduces and attacks primarily women. Thus, his vampiric transgressions operate mostly within a heterosexual framework. In addition, his promiscuity or penetration of multiple female throats, often in immediate succession, is ultimately punished because it violates the most significant precept of heteronormativity, monogamy. When he is not the Ripper, Stefan constantly represses his vampiric urges for profound ethical reasons and tortures himself by drinking animal blood. His commitment to one single woman (Elena), whose blood he occasionally ingests to keep his "bad" tendencies in check, suggests that he is eligible for domestication and can be accepted as a viable romantic hero only if he practices heterosexual monogamy. In the episode "Miss Mystic Falls," due to drinking human blood after a very long abstinence, Stefan falls off the wagon and compels Amber, one of Elena's peers, to become his blood slave. His relapse affects him so profoundly that he cannot face Elena. He tells Amber, "I've been drinking the human stuff, and it's really screwing with my head. She knows now! She wasn't supposed to find out! I didn't want her to find out! Now everything's ruined!" ("Miss Mystic Falls"). Ashamed and guilt-ridden, he almost commits suicide to prevent himself from betraying Elena or hurting other human beings again. It is also important to note here that Stefan the Ripper does embody, at one point in the show, a version of masculinity that is antithetical to traditional masculinity, especially in his ambiguously homoerotic rapport with Klaus Mikaelson, an Original vampire who craves Stefan's affection and pursues him quite aggressively. This alternative lifestyle, however, is temporary or never reaches its full potential, mainly because Stefan's "bad" or rebellious side seems to have been intended by the writers mostly as a plot device meant to allow the good boy turned bad to be eventually "cured" through heterosexual love. The "real" Stefan (i.e., the Stefan whose humanity is turned on) is not only defined by his adherence to heterosexuality, but he also abhors and rejects Klaus.

Like Stefan's bite, Damon's is as unrestrained and frequently accompanied by an increased sexual appetite, which he quenches by "compelling" or hypnotizing *only* female victims to perform favors for him, often sexual in nature. Throughout the first seasons in particular, Damon lives up to his bad boy reputation by frequently engaging in irresponsible "drinking" behaviors remorselessly. However, once he develops feelings for Elena and even before he is allowed access to her body, he ends up self-disciplining by consuming only human blood from blood packs or by entering monogamous relationships with women for blood benefits. Therefore, both Stefan and Damon are forced into adopting the heroine's (and by extension the audience's) standards, which end up being posited as the only viable reality that can truly redeem bad vampires. The implication is that there is great appeal or audience enjoyment in the process of turning bad boys into something that is safe or less alienating for the mainstream viewer.

While *True Blood* is more ambiguous than *The Vampire Diaries* in its depiction of sexuality, it actually follows a similar pattern in relation to Bill's and Eric's character development. Despite occasionally being involved in vague male-male relationships, both of them are generally presented as heterosexuals who may use their bodies for homosexual practices only to serve heteronormativity. For instance, in order to take revenge on Russell Edgington, a very, very bad homosexual vampire whose ambition is to annihilate humanity, Eric pretends to be sexually interested in his husband of 700 years, Talbot Angelis, whom he stakes to death during a sexual encounter (and importantly, before penetration occurs). As Darren Elliot-Smith argues, the trajectory of "gay" love is problematic in *True Blood*, due to its reliance on certain "gay archetypes and clichés" that construct the homosexual, embodied by Russell, as "psychotic" and "murderous" (149), as well as a threat that needs to be eliminated. In addition, while it could be argued that Eric shares an ambiguous intimacy with his sire-maker, Godric, it is primarily an intimacy that is never sexually interactive. Unlike Eric, Bill's character is almost entirely defined by his heterosexuality, most likely due to his status as the "good" vampire. As a human being, Bill strictly adhered to the heteronormative family model (i.e., marriage and children), while as a vampire, he cohabitated for many decades with his female maker and lover, Lorena. Even though he is once or twice depicted feeding on men in vampire nests and dark alleys, his transactions never go beyond the accepted boundaries of normative sexuality. Consequently, any ambiguity that could potentially destabilize the heteronormative standard is eventually annulled, or as Darren-Smith suggests, *True Blood* seems to be less concerned with the promotion of sexual diversity and more with the perpetuation of heteronormative assumptions (148–149). Thus, like *The Vampire Diaries*, *True Blood* ends up establishing and enforcing boundaries that are essential to recuperating the heteronormative order, which is essential for the domestication of the vampire with a bad-boy attitude or image.

As argued previously, the protagonists' ethnoracial identity is also essential for both identification and internalization. Especially *The Vampire Diaries* seems to cater primarily to a white audience. The majority of characters in the show—both human and vampire—are white; the exceptions are Bonnie, Elena's black friend, whose role is basically reduced to that of propping up her friend's whiteness, and the occasional black character who generally ends up dead within just a couple of episodes. In particular, the Salvatores' whiteness stands out as an asset in the show's portrayal of Connor, a black vampire whose practices are motivated by his hatred against the vampire's "ethnicity" and oppression. Fiercely devoted to his anti-vampire cause, which takes a Machiavellian form at times, he is depicted as a perpetrator of extreme, uncontrollable violence for which he is eventually punished. Interestingly enough, even though Stefan and Damon too commit acts of unspeakable violence, these acts go unnoticed, mainly because the two brothers act as protectors for the women they love, which in turn enable them to occupy a romantic position in relation to the female characters.

In *True Blood*, Bill capitalizes on his white history to facilitate his entrance into human society. As Victoria Amador remarks, as a former Confederate soldier and the son of a plantation owner, Bill "possesses two of the most important elements of the Southern social hierarchical structure—old blood and fine manners" (133). While this pedigree enables Sookie's grandmother to accept him as a potential suitor for her white and chaste granddaughter, it exacerbates the racial hostility between him and Sookie's black friends, Tara and Lafayette, who are decisively uncomfortable with his dual status as vampire and (former) white colonizer. Eric Northman, as his name suggests, is a millennium old Viking prince and inevitably linked to Odinism, which scholars argue was a religious movement propagating the "rhetoric of race war" and "white supremacy" (Kidd 225). These specific plot developments illustrate Kirkland's argument that the overall sympathetic portrayal of white vampires in contemporary television shows solely re-centers white privilege and white heroism, all the while re-inscribing the superiority of whiteness over non-whiteness (104–107). Consequently, a deconstruction of characters in both shows and the transactions that they engage in with one another suggest that under the bad boys' mask of rebellion one ultimately finds traditionalism.

In conclusion, the tension between heroes and antiheroes in contemporary cultural products as well as the romanticization of "bad boys" seems to be greatly informed by the mainstream's ideology of assimilation. What *The Vampire Diaries* and *True Blood* have in common is that they project two distinct typologies of masculinity within the romantic triangle: the "good" boy, whose goodness is defined in relation to his willingness to alter his values in order to gain membership into the mainstream and the "bad" boy who does not want to relinquish his "badness" but who is eventually rehabilitated into embracing the utopia of the dominant culture. As shown, these two typologies, despite their apparent oppositional relationship, end up serving the same purpose, that of reinforcing the legitimacy of the mainstream, which can preserve its centrality solely through exclusion or by forcing the margins either to assimilate or to perish. The "bad" vampire's identity in *The Vampire Diaries* and *True Blood*, despite its potential to challenge normativity is eventually posited as objectionable, while the human reality with its frailties and limitations is promoted as a more desirable alternative—one that the vampires who want to mainstream should aspire to. Consequently, the heroization of the "bad boy," at a fundamental level, does not challenge conservative ideologies but it is yet another mechanism through which the dominant culture maintains its hegemony.

Works Cited

Abercrombie, Nicholas, and Brian Longhurst. *Audiences: a Sociological Theory of Performance and Imagination*. London: Sage, 1998. Print.

Amador, Victoria. "Black and Whites, Trash and Good Country People in *True Blood*." Ed. Cherry Brigid. *True Blood: Investigating Vampires and Southern Gothic*. New York: I.B. Tauris, 2012. 122–38. Print.

Brumbaugh, Anna M. "How to Target Diverse Customers: An Advertising Typology and Prescriptions from Social Psychology." Ed. Steven S. Posavac. *Cracking the Code: Leveraging Consumer Psychology to Drive Profitability*. New York: Routledge, 2015. 66–89. Print.

"Cold Grey Light of Dawn." *The Vampire Diaries: Season Four*. Writ. Brian Young. Dir. Michael Ruscio. Warner Home Video, 2012. DVD.

"Disturbing Behavior." *The Vampire Diaries: Season Three*. Writ. Brian Young. Dir. Wendey Stanzler. Warner Home Video, 2012. DVD.

Elliot-Smith, Darren. "The Homosexual Vampire as a Metaphor for the Homosexual Vampire? True Blood Homonormativity and Assimilation." Ed. Cherry Brigid. *True Blood: Investigating Vampires and Southern Gothic*. New York: I.B. Tauris, 2012. 139–56. Print.

Faller, Bonnie. "'Vampire Diaries' Star Paul Wesley Bit a Fan! Find Out Why He'll Never Do It Again." *Hollywood Life*. 15 Feb. 2011. Web. 31 July 2016.

"Founder's Day." *The Vampire Diaries: Season One*. Writ. Brian Oh and Andrew Chambliss. Dir. Marcos Siega. Warner Home Video, 2010. DVD.

Kennedy, Randall. "Racial Passing." Ed. Kevin R. Johnson. *Mixed Race America and the Law: A Reader*. New York: New York University Press, 2003. 157–68. Print.

Kidd, Colin. *The Forging of Races: Race and Scripture in the Protestant Atlantic World*. New York: Cambridge University Press, 2006. Print.

Kirkland, Ewan. "Whiteness, Vampire and Humanity in Contemporary Film and Television." Ed. Deborah Mutch. *The Modern Vampire and Human Identity*. New York: Palgrave Macmillan, 2013. 93–110. Print.

"Miss Mystic Falls." *The Vampire Diaries: Season Three*. Writ. Brian Oh and Caroline Dries. Dir. Marcos Siega. Warner Home Video, 2010. DVD.
Reeser, Todd. *Masculinities in Theory: An Introduction*. Malden: Wiley-Blackwell, 2010. Print.
Schroeder, Jonathan. *Visual Consumption*. London: Routledge, 2002. Print.
True Blood. Seasons 1–7. HBO Home Entertainment, 2007–2013. DVD.

Deconstructing the Dichotomy That Is *House, M.D.*
A Carnivalesque Look at a Good Diagnostician/Bad Guy

Gillian Collie

In popular culture, characters with good guy/bad guy qualities abound. Protagonists may display poor behaviors that captivate their audiences. Gregory House, from the television series *House, M.D.*, is one such protagonist. Mikhail Bakhtin's popular culture theory of carnival exposes House as a protagonistic physician with good diagnostician/bad guy qualities. Bakhtin's parodic concepts of inside-out, feast of time, grotesque realism, and renewal trivialize House's qualities, and this trivialization illuminates the possibilities for deconstructing the dichotomy that is Dr. House.

Specific episodes of *House, M.D.* represent House as a brilliant, drug addicted, womanizing physician who is admired in spite of his irreverent behavior, because he solves problems. House rarely spends time with patients; he doesn't believe that patients are honest or reliable. He makes his subordinate interns see to patient care, and he receives most of the information on his patients from his subordinates and from patients' charts. House, in general, doesn't get along with the interns, other physicians, administration, or patients; however, he is tolerated and often indirectly admired because he is so adept at discovering patients' diseases without ever actually examining their bodies.

Because television shows like *House, M.D.* have a cultural influence on those who watch them, television can both reflect society and be reflected by it. And while there are no physicians who are really exactly like House, many aspects of House's personality and actions are predicated on a Bakhtinian trivialization of and reverence for actual physicians who may possess good diagnostician/bad guy qualities.

In a broad sense, culture is the study of social expressions in which groups of people partake. And ever since Karl Marx's "Capital," culture has taken on a political ideology. As Julie Rivkin and Michael Ryan state, "Culture is both a means of domination, of assuring the rule of one class or group over another, and a means of resistance to such domination, a way of articulating oppositional points of view to those in dominance" (1233). Whether or not this domination and subversion can exist simultaneously is questionable. But Bakhtin's theory of popular culture shows how House calls administrators' and physicians' authority into question.

Bakhtin's theory is based on a world where carnival represents denial of the hierarchy of the ruling class. Consecration of inequality refers to the acceptance of stratifications in social and economic classes as concrete and somehow *natural* when, in fact, they are abstract and socially constructed. In "Rabelais and His World," Bakhtin borrows Rabelais' suggestion that the festive carnival is one of the only places where consecration of inequality does not occur. In fact, "all were considered equal during carnival" ("Rabelais" 686). In opposition to the formal feast of the state where people are somber and the hierarchy is reinforced, carnival is predicated on laughter and is unofficially led by the popular sphere of the marketplace (686). But what conversations evoke equality during carnival? Inside out conversations would take place where rulers and laborers could dress in disguises and act as they liked. Feasts of time are conversations where people would attempt to become someone other than who they actually are. Conversations on grotesque realism would exaggerate bodily functions that all humans share in common. And, conversations on renewal would give a brief and momentary hope to all people as to the relativity of the hierarchy of authoritative rule. The world of *House, M.D.* includes carnivalistic notions of inside out, feast of time, grotesque realism, and renewal—all of which serve to demonstrate House's good diagnostician/bad guy characteristics.

House Appears and Talks as an Inside Out Patient

The idea of inside out is meant to turn the established hierarchy upside down. Bakhtin explains:

All the symbols of the carnival idiom are filled with this pathos of change and renewal, with the sense of the gay relativity of prevailing truths and authorities. We find here a characteristic logic, the peculiar logic of the "inside out," of the "turnabout," of a continual shifting from top to bottom, from front to rear, of numerous parodies and travesties, humiliations, profanations, comic crownings and uncrownings ["Rabelais" 687].

As Bakhtin elucidates, the acts of turning hierarchies inside out and iterating parodies towards prevailing truths and authorities are the acts of rebellion in which bad guys take part. In *House, M.D.*, House exposes his bad guy persona as he appears and talks as a patient, even though he is a physician. Additionally, House reprocesses the language of medical discourse in order to poke fun at it.

In the initial episode, "Everybody Lies," the first time we see House he is limping down the hall of the hospital, leaning on a cane for support. He is dressed in street clothes, with no white coat or name tag. Dr. Wilson, an oncologist and perhaps House's only friend, is walking beside him wearing a white coat. Wilson starts the carnivalesque conversation in the inside out world of *House, M.D.*:

> WILSON: "29 year old female, first seizure one month ago, lost the ability to speak, babbled like a baby, present deterioration of mental status."
> HOUSE: "See that? They all assume I'm a patient because of this cane."
> WILSON: "So, put on a white coat like the rest of us."
> HOUSE: "I don't want them to think that I'm a physician."
> WILSON: "You see why the administration might have a problem with that attitude?"
> HOUSE: "Um, people don't want a sick physician."
> WILSON: "That's fair enough; I don't like healthy patients."

The previous conversation illuminates Bakhtin's world of carnival where nothing is as it seems. House is preoccupied with his appearance even though he refuses to wear a white coat or a badge, and he acknowledges that he is a sick physician. He is a bad guy who defies the administration, although he is a brilliant diagnostician.

The dichotomy between the good diagnostician/bad guy in House becomes more apparent as the episode progresses, and it reflects the views of the popular culture about administration and patients. In addition to reflecting the views of the popular culture, the show also reflects popular culture's attitudes on the pomposity of medical discourse. The language of medical discourse is reprocessed in the series in order to poke fun at it. Bakhtin talks about the technique of the comic-parodic reprocessing of language:

> This usually parodic stylization of generic, professional and other strata of language is sometimes interrupted by the direct authorial word (usually as an expression of pathos, of Sentimental or idyllic sensibility), which directly embodies (without any refracting) semantic and axiological intentions of the author [sic] ["Discourse" 678].

When one applies Bakhtin's theory to *House, M.D.*, one finds that the parodic stylization of medical language is implemented by House. The dialogue that House uses illustrates his bad guy characteristics as he sarcastically spouts distain for patients, because he likes patients who cannot talk and therefore cannot communicate. Also, House and Wilson banter back and forth with one another about practicing medicine without actually knowing anything, and complementing one another in a parody of pretense about being simple country physicians. The heteroglossic discourse is evident because medical discourse is mixed with social discourse. Even though House and his colleagues use very scientific medical words, usually to indicate diagnoses, most of the conversations use social discourse. In fact the social discourse is very informal, using as many contractions and slang terms as possible. Words such as "I'm" and "don't" are used very frequently, and words such as "guy" or "doctor" are used as often as medical discourse is used. This reprocessing of language adds to the carnivalesque nature of an upside down house of medicine that houses a good diagnostician/bad guy.

House is aware that Wilson is lying to him about having a familial connection to the patient because House suspects everyone of lying. House is also disgusted with HMOs, as are many real life physicians and patients who may be watching the series. Yet House must be respected as a physician in some ways, because Wilson is consulting with House and because House has a team of overqualified physicians working for him. Wilson knows that House is more interested in the abnormal physiology of disease that coroners get to witness than he is with living, healthy patients. And House, as well as being chronically disabled, is a drug addict; none of these attributes is what most people think of when they think of revered physicians. Instead, House fits the profile of a bad guy.

In the next scene, House is with his team: Drs. Foreman, Chase, and Cameron. They are gathered around a white board that they use in order to assess specific diagnoses, and House declares that everybody lies. House would rather look at supposedly objective science than rely on patients' histories. Because House believes that everyone lies, he is much more interested in uncovering the nature of the illness than he is in treating the patient. House's views are the exact same views that actual physicians are trying to figure out how to get rid of, yet House is a bad guy because he is modeling them to his team.

While House waits for the elevator, the hospital administrator, Dr. Cuddy, joins him. They engage in sexual banter. House is irreverent to her because he refuses to come to her office and he continually dismisses requests for consults from her. The scene pictures Cuddy as an articulate physician who employs more formal speech than House. Her words are all business. She wears a business suit as opposed to House's wrinkled shirt and jeans. Both Cuddy and House acknowledge House's paradoxical value to patients as a good diagnostician and simultaneous lack of adherence to the rules as a bad guy. House's contradictory nature may reflect the views of the popular sphere that wants to believe that it is highly valued for its work while it is irreverent towards the rules. In this scene, there is a glaring juxtaposition linking the harsh words and intimate glances between these two individuals. House and Cuddy are standing really close to one another, and there is obviously sexual tension between them. House's charisma must compensate for his caustic retorts, otherwise why would Cuddy be attracted to him?

As the episode continues, it becomes clear that the patient, Rebecca, does not know who her physician actually is. Although House may accurately diagnose Rebecca's disease he does a poor job of interacting with her, or else she would know who he was.

In the next scenes, there are many examples of House's irreverent behavior: House keeps popping pills; Cuddy takes away House's television privileges until House agrees to clinic duty; Rebecca sees House loitering outside her room, bent over his cane; and, House pressures Foreman to break into Rebecca's house to see about possible contaminants that may be making her sick. Overusing drugs, avoidance of work duties, avoidance of direct contact with patients, and burglary are bad guy actions—but are nevertheless accepted from House because he is such a good diagnostician.

House watches soap operas and reads tabloids. The fact that he does so indicates actions that are in direct contrast to what people often think typical physicians do with their time. Typical physicians are busy treating patients, not being engrossed in the basest forms of popular culture. This is further evidence of Bakhtin's carnival where House appears and acts as a bad guy who participates in mediocre pastimes.

House surmises that Rebecca may have tapeworms in her brain from an undercooked ham. House's team decides to x-ray Rebecca's leg and they find tapeworm eggs, so Rebecca consents to treatment. The episode ends when Rebecca is treated and cured with two pills, though the exact treatment that Rebecca takes is not specified. Rebecca tells Cuddy that House is the best physician ever. Cuddy rolls her eyes in disgust at the patient's idealistic assumption (Singer). House's reality is that although he is a good diagnostician, he is also a bad guy.

And even though Bakhtin's theory of carnival illuminates the contrary nature of humans, it also allows humans the possibility of reinventing themselves.

House Attempts to Reinvent Himself by Tricking Time

Tricking time is an important aspect of carnival. Bakhtin asserts that "carnival was the true feast of time, the feast of becoming, change, and renewal. It was hostile to all that was immortalized and completed"

("Rabelais" 686). Can House overcome his bad guy persona during the true feast of time? House is a drug addicted womanizer but he attempts to trick time—and perhaps himself—by discontinuing narcotics and engaging in a monogamous relationship with Cuddy.

Wilson and House are living together, but Wilson ejects House from his apartment. Additionally House's domain is occupied by Alvarez, an illegal alien who sells off House's most sacred possessions. House's narcotic addition reaches its peak when House almost kills one of his patients, and starts to hallucinate about an imaginary relationship with Cuddy. When House forces himself to spend time at Mayfield Psychiatric Hospital, he stops taking narcotics and engages in counseling with Dr. Nolan. Methadone treatment helps House to walk on his deformed leg without pain. "Baggage" elucidates House and Nolan's conversations. Even though House is rude and irreverent to Nolan, House comprehends that Nolan is as brutally honest as House is himself. Through his discussions with Nolan, House realizes that people all around him are happy, moving in together, and living their lives without regret.

Ironically, House begins to feel that he may actually belong in the world just as he is displaced from his own house. House is a metaphor; it may be a metaphor for a constructed space that is not a home. In psychiatric medicine, a "nut house" is slang for an insane asylum. In this case, *House, M.D.* may be a metaphor for a nutty house of medicine with an inside out physician; or patients may simply house the diseases that House likes to diagnose. Whatever the case, House gains hope of letting go of his bad guy personality as he is dislodged from his house and is located in a psychiatric hospital.

The combination of displacement, a sober disposition, and therapy allows House to envision the good guy kind of life that he wants; House wants a monogamous relationship with Cuddy. "Now What?" is the episode where House and Cuddy consummate their relationship for the first time. House appears to be receptive to a real relationship, but even the first discussion between House and Cuddy is a prediction of impending destruction. When House asks Cuddy, "Now what?" Cuddy assumes that the sexual relationship is already over. But House was just wondering what the two of them should do for the rest of the day. The episode shows that House, at least temporarily, has the ability to nurture Cuddy. House makes Cuddy breakfast, pours her a bath, and gives her a massage. As the episode progresses, House and Cuddy have sex again, and each is wondering if the other wants to keep their sexual companionship a secret from the rest of the world. House, almost immediately, thinks that there is a problem and that the relationship will not last. He is afraid that Cuddy will hate him for the drug addict that he is. Cuddy says that she thinks House is afraid to be happy because, for him, happiness never lasts. At the episode's closing, House declares his love for Cuddy.

The banter that accompanies House and Cuddy's sexual escapades foreshadows the decline and eventual dissolution of their relationship. And the affiliation dissolves as House restarts his bad guy traits. When House resumes his use of narcotics, he hides it from Cuddy. The dishonesty and secrecy are more than the connection can handle.

"Moving On" is the episode where House and Cuddy acknowledge the dissolution of their relationship. House attempts to remove tumors from his own leg, tumors that result from an experimental drug that House was taking. Cuddy asks House how he feels, and of course House flippantly answers in reference to his leg pain instead of according to his failed connection with Cuddy. Finally House concedes, "I feel hurt" which is a response to both his mutilated leg and his failed involvement with Cuddy. As House returns home he discovers that Cuddy left her hair brush behind, and decides to return it to her. House gets in his car and drives it straight into Cuddy's living room, where she is entertaining family and friends. Casually, House gets out of his car and gently places Cuddy's hairbrush in her hand and then leaves. The car remains inside Cuddy's House.

House disrupts Cuddy's relationships and home; he does so literally with his car and metaphorically with his influence on her. This bad boy purposefully ruins his relationship with Cuddy beyond repair. Because he is afraid that he is not good enough, even as he attempts to shed his bad guy persona, House vandalizes Cuddy's house so that she cannot possibly want him back. And House's violent actions allow him to absolve himself of having tried and failed at being a good guy. House's prevailing truth is that he is a bad guy and although he temporarily tries to liberate himself from his truth and trick time, House ends up back on drugs and without a significant companion.

Even though carnival allows for the possibility of liberation from prevailing truths, it simultaneously exposes ugly realities as feasts of time expire. In addition, carnival unearths the ugly realities of bodily functions and malfunctions.

House Marvels at the Grotesque Realism of His Patients

Grotesque realism is another of Bakhtin's important carnivalesque concepts. In grotesque realism "[t]he material bodily principle is contained not in the biological individual, not in the bourgeois ego, but in the people, a people who are continually growing and renewed. This [grotesque realism] is why all that is bodily becomes grandiose, exaggerated, immeasurable" ("Rabelais" 688). *House, M.D.* demonstrates grotesque realism through trivialization of diseases exposed by a good diagnostician/bad guy, under the pretense of attendance to the biological individual.

One episode of *House, M.D.* that conveys grotesque realism is "No Reason." A grotesque tongue calls attention to the public sphere's inability to speak. The episode opens with a full screen view of a swollen tongue. Oddly, House actually agrees to see this patient. He thinks it's hilarious to try and listen to the afflicted patient talk. The tongue is five times its normal size. The scene shows a huge biopsy needle going into the enormous red, swollen, useless tongue. This scene draws attention to all human tongues. We cannot help but grab our own tongues in pain and realize that the tongue on the screen becomes a collective tongue that cannot speak utterances. Tongues may, at times, be beautiful and sensual, but, at others, be grotesquely deformed so as to expose their vileness. House accurately diagnoses the tongue's ailment as he simultaneously disparages the patient's predicament, reflecting his erratically good and bad actions.

Another instance of grotesque realism in *House, M.D.* focuses on a beautiful dancer who loses her skin. A prima ballerina becomes grandiosely deformed in "Under My Skin." Because of a rare infection, the woman's skin sloughs off faster than the artificial protective skin can replace it. The patient loses eighty percent of her skin. She turns from a beautiful ballerina into a red, oozing, deformed monster. House declares that the patient has toxic epidermal necrolysis, and admits that he treated her with an antibiotic without first confirming that she had the infection. The patient does not seem aware that she is ugly; she only wants to dance again. The situation is ironic because nobody wants to watch an ugly ballerina perform. We ache for the ballerina and her predicament, and we are in on the collective reality that she is too hideous to perform in public. And when we are unable to gaze upon even the most beautiful of humans, the situation is grotesque. House exposes his bad guy manner when he openly declares that the ballerina is repugnant and that he treats her without following the appropriate protocol.

An additional episode of *House, M.D.* deals with the grotesque realism of crucifixion. And people may, at times, feel crucified by the public. In "Small Sacrifices" a man nails himself to a cross once a year, as a deal with God to keep his daughter healthy. Four years earlier, the father's daughter had cancer. Three weeks after his initial self crucifixion, the father's daughter was cancer free. Now the father repeats the process every year in order to keep God satisfied. The screen shows the nails being pounded slowly and decisively into the patient's flesh. Several men raise the burdened crucifix. The scene portrays intensely hot sun juxtaposed to blood spilling out of the father's mouth, hands, and feet. House is only interested in the father because of his need to self crucify, not because of his wounds. House suspects that the father is atoning for something bad that he has done in the past, and he does not believe the father's story about his daughter's cancer. And House calls the patient crazy and stupid, and claims that ritual is what people do when they run out of rational. At this point, a bad guy trait of House is exposed because good physicians do not judge patents' ritualistic faith. When the public sphere thinks about crucifixion, we immediately consider Christ. But in *House, M.D.*, we experience a grotesquely twisted scene of carnival that forces us to empathize with the father by contemplating our own crucifixion. Even if we cannot contemplate our own crucifixion, we can, hopefully, empathize with the father. At the center of the public sphere are our children, and many of us would suffer a crucifixion in order to atone for the sins of our children and give them everlasting life. Many of us do indeed make many sacrifices for the benefit of our children.

Further, grotesque realism is present in an episode of *House, M.D.* where House exaggerates a story in order to shock children. "Two Stories" is a tale about a man who coughs up a lung within a story about House talking to a group of grade school students on their career day. Cuddy forces House to talk to school children about his profession. Perhaps in order to punish Cuddy or because of his irreverence for the situation, House tells the story in an extremely grotesque manner. House knows he is inappropriate because he does not use his real name. Instead, House fraudulently calls himself Dr. Hourani when addressing the class of kids. House is dishonest and graphic in telling the story of the man with a chunk of his own lung in his hands. As House recalls the story, the patient coughs and red blood and white tissue come up in the patient's mouth. The patient supposedly coughs his lung all over the nurse and physicians that are in the room, except that the tissue is not really lung tissue. The physicians in the room ignore the fact that the patient is coughing up blood because they are flirting with a new nurse. The other physicians finally notice what is happening to the patient when a huge hunk of dark red tissue appears in the patient's hand. A large portion of hard dead tissue takes up the entire screen. As House reveals the story to the fifth graders, he claims that none of the other physicians could figure out what was wrong with the patient but it is he who realizes that the patient must have inhaled a piece of food down the wrong pipe. House does exploratory surgery on the patient and finds a single pea in his airway. House saves the patient by removing the pea. The scene is doubly grotesque because the coughing up of a lung is exaggerated and untruthful, and the story's inappropriate nature is witnessed through the collectively innocent eyes of school children. Furthermore, even though the children idolize House as a heroic physician for an embellished story of heroism, House is a bad guy because he is unable to take accountability for his lies.

But perhaps House has witnessed enough grotesqueness to make a space for renewal.

House Undergoes a Renewal

Renewal is a fourth of Bakhtin's important carnivalesque concepts. During the comedic ritual of carnival, the common sphere devises symbols of renewal in order to provide temporary hope that the authoritative hierarchy is not so rigid or unchangeable. Remember that Bakhtin asserts, "All the symbols of the carnival idiom are filled with this pathos of change and renewal, with the sense of the gay relativity of the prevailing truths and authorities" ("Rabelais" 687). But can House, the good diagnostician/bad guy, become filled with change and renewal?

House does indeed become a renewed character in the series finale. In "Everybody Dies," House is in a burning building trying to decide whether he wants to escape or to become enveloped in the flames. As he meditates on his decision, House sees ghosts from his past. He sees ghosts of previous team members who have died, team members who are still alive, and his ex-wife. He thinks many thoughts: death is not interesting, but life isn't interesting anymore. Then suddenly, House falls through the second floor of the burning building.

Wilson the oncologist is, ironically, dying of cancer. House doesn't want to lose his last days with Wilson, but it is House who is in the burning building.

Wilson and Foreman see House through the window of the burning building, and then the building explodes. The two witnesses confirm that House died in the fire.

Wilson's eulogy praises, blames, and abuses House:

> Gregory House saved lives, he was a healer.... House was an ass. He mocked anyone, patients, coworkers, dwindling friends, anyone who didn't measure up to his insane ideals of integrity. He claimed to be on some heroic quest for the truth, but the truth is he was a bitter jerk who liked making people miserable.

Wilson gives the eulogy in a particular Bakhtinian carnivalesque parody where encomium is followed by invective. Wilson starts by calling House a savior and a healer, and then quickly moves on to calling him an ass and a bitter jerk. Bakhtin explains the parodic style:

> The passing from excessive praise to excessive invective is characteristic, and the change from the one to the other is perfectly legitimate. Praise and abuse are, so to speak, the two sides of the same coin. If the right side is praise, the wrong side is abuse, and vice versa.... The praise, as we have said, is ironic and ambivalent. It is on the brink of abuse; the one leads to the other, and it is impossible to draw the line between them ["Rabelais" 690].

The fact that abuses are directed towards House is particularly funny, given the circumstances that this is House's eulogy, and eulogies are supposed to be encomiums. Further, this abusive language serves an additional function. Abusive language is a special type of degradation that leads to regeneration. Bakhtin argues, "Degradation digs a bodily grave for a new birth; it has not only a destructive, negative aspect, but also a regenerating one" ("Rabelais" 688). And House's regeneration comes from his resurrection in the fire.

Wilson figures out where House is hiding, and Wilson finds House at Wilson's house. House acknowledges that he got out through the back of the burning building, and switched his dental records with an unknown victim that was burned in the fire. House's manipulation of the victim and his records once again demonstrates House's bad guy side.

As the series closes, House and Wilson drive their motorcycles down the road together. Wilson and House both leave medicine, Wilson in order to live out his last months and House in order to reinvent himself. House's regeneration is a rebirth. Like the Greek phoenix, House burns up in the flames and is reborn from the ashes. House takes the initiative to change his circumstances permanently because he abandons his persona as a good diagnostician as he rides his bike down the road with Wilson. And perhaps, House may be able to abandon his persona as a bad guy as well.

House Transforms from a Good Diagnostician/Bad Guy to a Human Guy

House burns his old life so that he may start a new life, his only objective being companionship for Wilson during Wilson's final days. Wilson overtly states that House is on a heroic quest. But throughout the series, House is not a hero. He is a protagonist who, like all humans, has his good qualities and bad qualities. House is a gifted diagnostician who is a bad guy. Is he a bad guy because he is uncomfortable with his brilliance? Is he a bad guy because he is a misanthrope? Is he a bad guy because he is awkward with or afraid of social interactions? These questions are never directly answered; however, in the last scene of the series House makes a transformation. And, he does it by literally walking away from his old life. Whether or not House escapes his good diagnostician/bad guy persona is, however, another question.

Dichotomous characteristics like House's are often described in terms of binary oppositions. As with all binaries, we must deconstruct the binary of good diagnostician /bad guy in House's character in order to dismantle prejudicial preconceptions of these two signs; this deconstruction becomes increasingly important as we replace the binary good diagnostician/bad guy with human guy. After all, humans possess many and various dichotomies within their character and these dichotomies elucidate their humanity—if one binary is not preferred over the other and both are accepted.

One of the few selfless acts that House performs in *House, M.D.* is to spend time with Wilson before he dies; however, House will probably continue to be arrogant and selfish in his new life. But maybe leaving his position as a brilliant diagnostician will help House to temper his bad guy characteristics. And perhaps, House will be able to accept his dichotomously human nature.

WORKS CITED

Bakhtin, Mikhail. "Discourse in the Novel." Rivkin and Ryan 674–85.
_____. "Rabelais and His World." Rivkin and Ryan 686–92.
Rivkin, Julie, and Michael Ryan, eds. *Literary Theory: An Anthology.* 2d ed. Malden: Blackwell, 2004. Print.
Shore, David, dir. "Everybody Dies." *House, M.D.* Perf.: Hugh Laurie, Omar Epps, and Peter Jacobson. 8.22 2012. Television.
_____. "No Reason." *House, M.D.* Perf.: Hugh Laurie, Lisa Edelstein, and Omar Epps. 2.24 2005. Television.
Singer, Bryan, dir. "Everybody Lies." *House, M.D.* Perf.: Hugh Laurie, Lisa Edelstein, and Omar Epps. 1.1 2004. Television.
Straiton, David, dir. "Baggage." *House, M.D.* Perf.: Hugh Laurie, Lisa Edelstein, and Omar Epps. 5.10.2010. Television.
_____. "Under My Skin." *House, M.D.* Perf.: Hugh Laurie, Lisa Edelstein, and Omar Epps. 5.16 2008. Television.
Yaitanes, Greg, dir. "Moving On." *House, M.D.* Perf.: Hugh Laurie, Lisa Edelstein, and Omar Epps. 5.23.2011. Television.
_____. "Now What?" *House, M.D.* Perf.: Hugh Laurie, Lisa Edelstein, and Omar Epps. 9.20.2010. Television.
_____. "Small Sacrifices." *House, M.D.* Perf.: Hugh Laurie, Lisa Edelstein, and Omar Epps. 7.8 2010. Television.
_____. "Two Stories." *House, M.D.* Perf.: Hugh Laurie, Lisa Edelstein, and Omar Epps. 7.13 2010. Television.

About the Contributors

Hilarie **Ashton** is a Ph.D. candidate in English at the CUNY Graduate Center, where she focuses on American studies and composition. Her work has appeared in the *South Atlantic Review* and the *Journal of Popular Culture*, among other venues.

Alissa **Burger** is an assistant professor of English and director of Writing Across the Curriculum at Culver-Stockton College. She is the author of *The Wizard of Oz as American Myth* (McFarland, 2011).

Jack **Clarke** is a Ph.D. candidate at Monash University in the School of Media, Film and Journalism. His research areas include online television audiences, contemporary masculinity, and social media analysis.

Gillian **Collie** received a Ph.D. in English from Indiana University of Pennsylvania in 2015. Her research focuses on discourse communities in rhetoric and composition.

Ana G. **Gal** is an English instructor at the University of Memphis. She researches 19th-century British literature, the Gothic, and cultural studies, as well as Anglo-American representations of monstrosity and the politics of performativity.

Tamara **Girardi** is an assistant professor of English at Harrisburg Area Community College. She also writes young adult fiction.

Robert **LaRue** is an assistant professor in the Department of English at Moravian College. His work addresses the presence and roles of politics within postcolonial queer expressions.

Tirza I. **Leader** is an assistant professor of psychology at Georgia Gwinnett College. She uses methodologies from anthropology and sociology to study problems associated with prejudice and utilizes interdisciplinary solutions to research problems.

Stephanie **Lim** is a Ph.D. candidate in drama and theatre at University of California, Irvine. Her work has investigated ecocriticism in *The Little Shop of Horrors*, cross-cultural relationships in *The Book of Mormon*, and representations of minority cultures in Los Angeles productions of *Pippin*.

About the Contributors

Richard L. **Mehrenberg** is an assistant professor of special education at Millersville University. His research interests include coteaching, generational theory, and the needs of first-year teachers.

Darcy **Mullen** is a 2017 Postdoctoral Marion L. Brittain Fellow at the Georgia Institute of Technology. Studies. Her publications explore the use of "local" as a tool for cartography in the local food movement, the politics of place and tourism in Myanmar, and pedagogies in disability studies.

Sotiris **Petridis** is a postdoctoral researcher at Aristotle University in Thessaloniki, Greece where he teaches film theory and screenwriting. His research interests include film and television genres, audiovisual rights and copyright laws, viral marketing, and the new ways of film and television promotion.

Abigail G. **Scheg** is a course mentor for General Education Composition at Western Governors University. She is the author or editor of a number of texts including *Bullying in Popular Culture* (McFarland, 2015).

Annette **Schimmelpfennig** is a research assistant at the University of Cologne, where she teaches postmodern and horror fiction. She has published on portrayals of gender in contemporary TV shows, fairytales, and Christopher Nolan's The Dark Knight Trilogy.

Hannah **Swamidoss**, an independent scholar, has a Ph.D. from the University of Texas at Dallas and has published articles on various children's books. Her interests also include children's film and television shows.

Don **Tresca** specializes in 20th-century American literature and film studies. He has previously published essays on a wide variety of pop culture subjects, including the works of Joss Whedon, J.K. Rowling, Stephen King, and Clint Eastwood.

Lloyd Isaac **Vayo** is a lecturer in cultural studies and comparative literature at the University of Minnesota–Twin Cities. He has published on 9/11 and terrorism, popular music, and literature, and is coeditor of *Terror and the Cinematic Sublime* (McFarland, 2013).

Caroline **Womack** teaches English at Frisco High School in Texas. She earned her MA in English Literature from the University of Leeds and has presented on Medieval and Renaissance literature at several conferences.

Jacinta **Yanders** is a Ph.D. candidate in the Department of English at Ohio State University. She researches representations of race, gender, and sexuality in media as well as contemporary media trends.

Index

Agents of Shield 136–137, 143–145
All My Children 162
American quality television 74
The Americans 7
Angel 136–140
Archetypes 33, 109
The Avengers 138

Backstrom 149–161
Bakhtin, Mikhail 198–109
Better Call Saul 33
Billions 33
Black Lives Matter 49
Boardwalk Empire 33
Bonnie and Clyde 34
Breaking Bad 8, 84–95, 150
Brody, Sergeant Nicholas 174–184
Brown, Michael 19
Buffy the Vampire Slayer 60, 140

Camelot 109, 111
Chappelle, Dave 55
Chipping, Mr. "Chips" 84–85
complexity, linguistic 65
Conrad, Joseph 123
Corinthos, Sonny 162–172

Damages 77
Dexter 33, 150
Dollhouse 136–137

Easy Rider 34
Empire 96–107

Feminism 60, 77
Firefly 136
The Flash 22
Flesch Reading Ease 65

Gang Related 7
General Hospital 162–172
The Godfather 166
The Good Wife 77

Grey's Anatomy 77
Grimes, Rick 1, 123–134

Hannibal 33
HBO 33, 75–76
Homeland 33, 74, 78–79, 174–184
House 198–109
House of Cards 33, 74, 79–80
How to Get Away with Murder 8, 74, 80–82

It's Always Sunny in Philadelphia 55

Key, Keegan-Michael 48
Key & Peele 48–58
Kirkman, Robert 1

Law & Order 8
leadership 123–134
LGBT 97–99
Lost 75

Mad Men 33
Marx, Karl 199
memes 33, 38–44
Merlin 109–122
Montana, Tony 34
Le Morte d'Arthur 109–112, 116

National Board Certification 86
Negan 1
Netflix 55
Nielson Rating 60

Obama, Barack 48–58
One Life to Live 162
Orange Is the New Black 7, 14
Other 16
Oz 7, 10

Peele, Jordan 48
Pinkman, Jesse 86–87, 90
Prison Break 75

213

queer studies 99–101, 103

Rice, Tamir 19

Salvatore, Damon 2, 187 195
Scandal 7, 9, 77
Scarface 34, 84
The Shield 33, 151
Sons of Anarchy 7, 33
The Sopranos 33, 34, 150
stereotypes 129–130
superheroes 23

The Terminator 63
True Blood 186–196

The Vampire Diaries 2, 186–196
violence 61–64; 69–70

The Walking Dead 1, 123–134
Wells, Harrison 22–28
Whedon, Joss 136–145
White, Skyler 35–36, 43–46, 92–93
White, Walter 35–46, 84–95, 150
Whitman, Walt 96
The Wire 151
World War II 23

YouTube 37

www.ingramcontent.com/pod-product-compliance
Ingram Content Group UK Ltd.
Pitfield, Milton Keynes, MK11 3LW, UK
UKHW041958140426
5217IPUK00015B/863